PRACTICAL CONIC SECTIONS

J. W. DOWNS

DALE SEYMOUR PUBLICATIONS

Managing Editor: Michael Kane
Project Editor: Priscilla Cox Samii
Production Manager: Janet Yearian
Production Coordinator: Leanne Collins
Design Manager: Jeff Kelly
Cover and Text Design: Lisa Raine

Photo Credit: cover and page v, The Bettmann Archive

This book is published by Dale Seymour Publications, an imprint of The Alternative Publishing Group of Addison-Wesley.

Copyright © 1993 by Dale Seymour Publications. All rights reserved. No part of this publication may be reproduced in any form or by any means without the prior written permission of the publisher. Printed in the United States of America.

Order Number DS21201
ISBN 0-86651-628-X

4 5 6 7 8 9 10-MA-00 99 98 97

CONTENTS

PREFACE		vi
INTRODUCTION		vii
CHAPTER ONE	Deriving Ellipses	1
CHAPTER TWO	Deriving Hyperbolas	10
CHAPTER THREE	Deriving Parabolas	17
CHAPTER FOUR	The Directing Circle	25
	Constructing an Ellipse	26
	Constructing a Hyperbola	28
	Constructing a Parabola	30
	Points of Tangency	31
	Comparing Conic Curves	34
CHAPTER FIVE	Reflective Properties of Solid Conic Curves	36
	Ellipsoids	36
	Hyperboloids	41
	Paraboloids	43
CHAPTER SIX	Compound Reflectors	46
	Cassegrain Systems	46
	Gregorian Systems	48
	Using the Cassegrain Principle	48
	Compound Reflector Design	50
CHAPTER SEVEN	Eccentricity	55
	Special Cases	56
CHAPTER EIGHT	For Practical Purposes	60
	Telescopes	60
	Sound Applications	60
	Variations	62
	Acoustical Applications	65
CHAPTER NINE	Unusual Properties of Cones and Conic Curves	68
	An Interesting Proof	68
	Cones	71
	Ellipses	74
	Kepler's Laws	75
	Orbital Velocities	82
	Elliptical Gears	83
	Sketching Conics from a Formula	84
	Conclusion	89
APPENDIX A	Nonparabolic Reflectors for Antennas	90
APPENDIX B	Computer Program for Generating Conic Curves	95
BIBLIOGRAPHY		100

"Do not disturb my circles." These were the last words of Archimedes, who was killed in 212 B.C. by Roman soldiers as he worked geometry problems in the sand.

A tracing of Kepler's proof that equal areas were swept in equal time in an elliptical orbit.

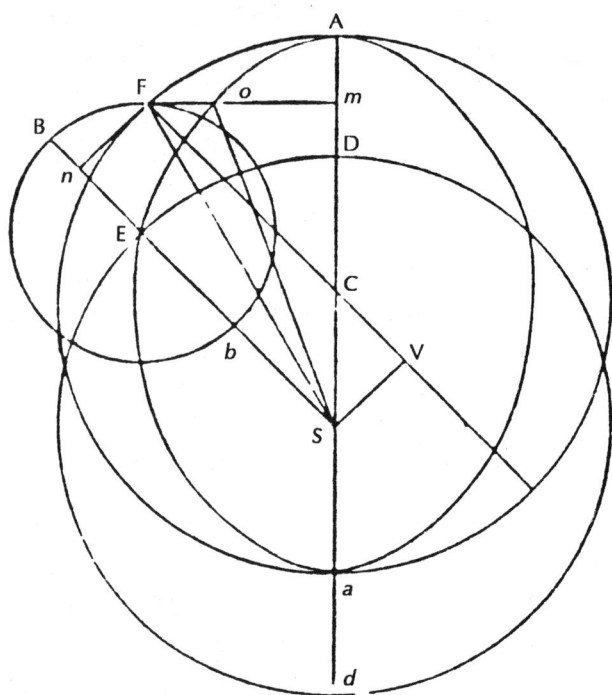

"The geometrical mind is not so closely bound to geometry that it cannot be drawn aside and transferred to other departments of knowlege. A work of morality, politics, criticism, perhaps even eloquence will be more elegant, other things being equal, if it is shaped by the hand of geometry."

Bernard le Bovier de Fontenelle
Préface sur l'Utilité des Mathématiques et la Physique
(1729)

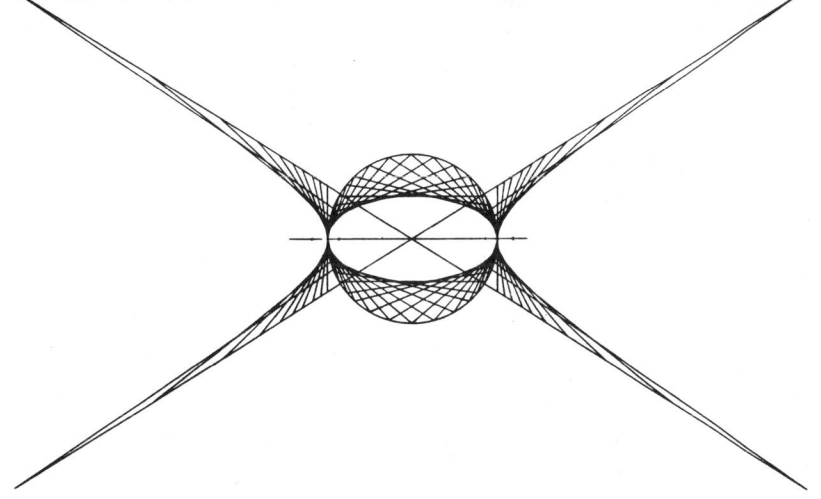

Ellipse and hyperbola defined by tangents and directing circle.

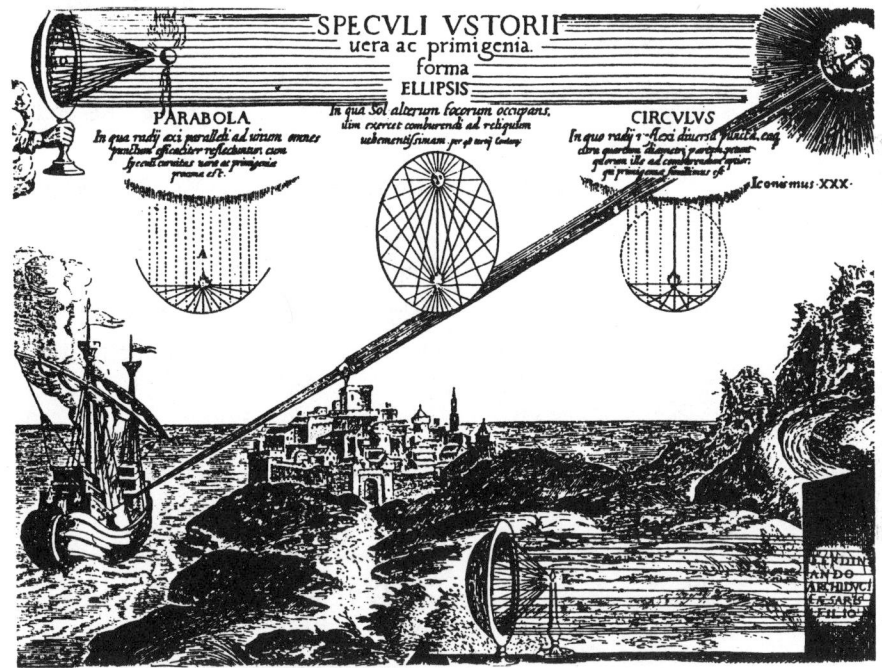

Reflective properties of conic sections have been known for two thousand years. This seventeenth-century engraving by Kircher illustrates Archimedes' legendary burning mirrors.

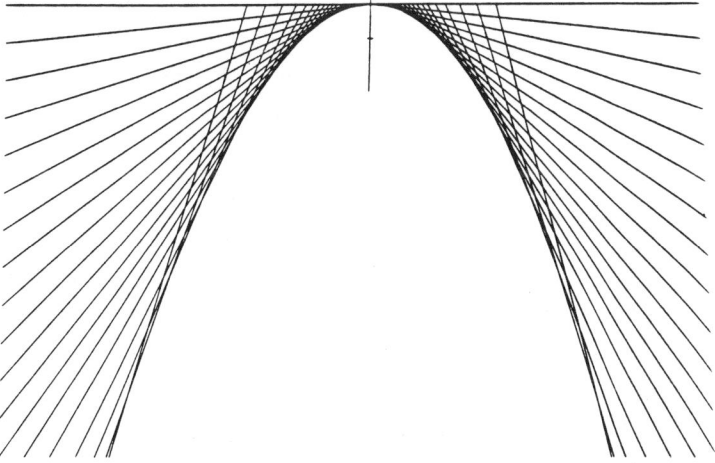

Parabola defined by its tangents.

PREFACE

One of the fortunate aspects of pure geometry is that there is no need to depend on references to authorities on the subject to substantiate claims, as is the case in the less precise sciences. Geometric proofs are completely self-evident. There are several items in this book that are original, to the best of my knowledge; so citing an ancient (or modern) authority is impossible. These items have been included because they stand up to the rigors of geometric proof and may be repeated by anyone.

A moderate knowledge of plane geometry is helpful in understanding this book, but no knowledge of mathematics (algebra, etc.) is required. The objective is to give a conceptual understanding of conic curves—something that is almost totally lost when these curves are reduced to mathematical formulas.

Every method of deriving conic curves that I have found has been included in Chapters 1–3. This is not to say that they are the only methods known. Not all construction methods described are practical, but they have been included because they can be made to work. For example, it is doubtful that anyone wishing to generate one of these curves would actually make a cone and slice it at the appropriate place and angle to yield the desired curve. It is possible, yet impractical, to do so. Many of the construction methods are shown here for the record and may be skimmed over without impairing the practical approach to the understanding of these curves.

The final chapter is a compilation of unique mechanical and reflective properties of cones and conic curves. Additional information would be welcomed in the event this book should go into a second printing.

J. W. Downs
Santa Clara, California

INTRODUCTION

"In conics I can floor peculiarities parabolous."
Gilbert and Sullivan, *Pirates of Penzance*

Since the time of Alexander the Great, there has been an interest in cones and conic sections. Until the time of Apollonius of Perga (the Great Geometer) in the third century B.C., it was thought that ellipses were taken from acute cones, parabolas from right angle cones, and hyperbolas from obtuse cones, as shown in Figure 0-1.

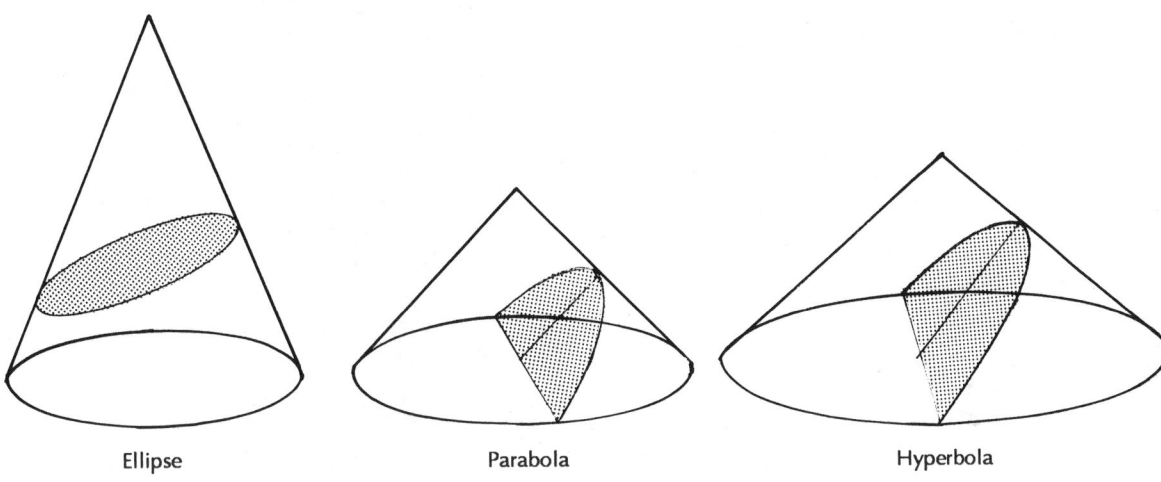

FIGURE 0-1

Apollonius demonstrated that all conic curves could be obtained by sectioning any cone, acute or obtuse, simply by changing the angle at which the plane intersects the cone. Although he wrote eight books (chapters) on conics and named ellipses, parabolas, and hyperbolas, he and the others of his time missed the most important attributes of conic sections, and no thought was given to any practical use of this unique family of curves.

Figure 0-2 shows a pair of cones joined at the apex. Lines AB, AC, AC', and AD represent planes pivoting at point A and cutting one or both cones. As the plane cuts through the cones, it will encounter ellipses within the sector where it passes completely through the cone (AB). When the plane is parallel to one side of the cone (*AC* or *AC'*), the plane will no longer pass completely through the cone but can be extended forever. At this point the figure is a parabola. Beyond this angle the plane will never leave the cone (*AD*), and it will intersect the other cone, generating the two halves of a hyperbola.

FIGURE 0-2

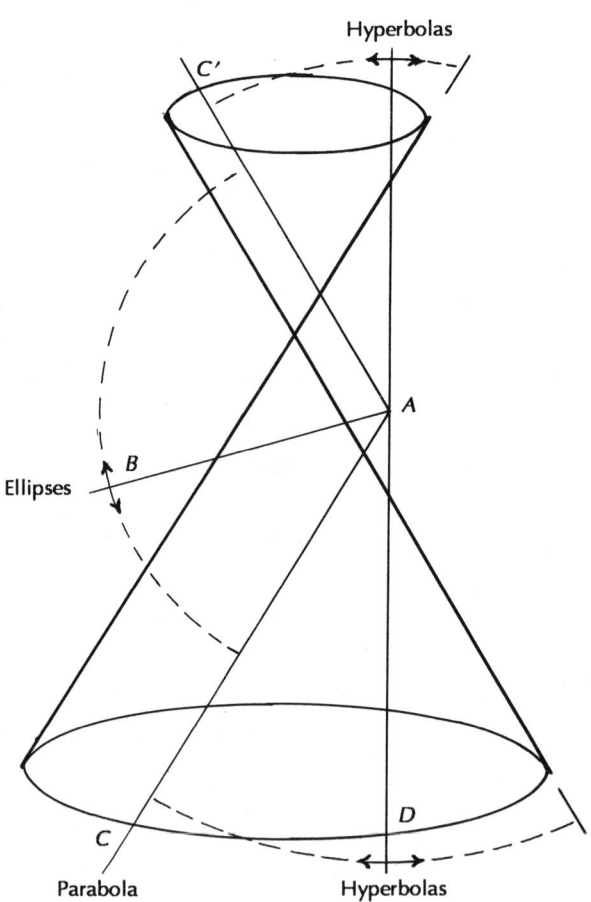

Anthemius of Thrales, the architect of the Church of the Holy Faith in Istanbul in the sixth century, wrote a book called *Concerning Remarkable Mechanical Devices,* in which he discussed burning mirrors and the reflective properties of solid conic curves.

Archimedes probably experimented with parabolic reflectors, although there is no reference to this in his surviving works. The story that he devised burning mirrors to set the Roman ships on fire can only be legend unless the captains of the Roman ships were uncommonly accommodating. The focal length of the reflector (or lens) cannot be more than approximately twice its diameter to gather enough solar energy to start a fire. The Roman ships would have to have been docked within a few feet of the mirrors and carefully positioned with relation to the sun for burning mirrors to work.

As the science of astronomy developed more accurate ways of measuring planetary motion, it became evident that the planets were not moving in circular orbits around the sun. Kepler discovered that planetary motion was elliptical, with the sun at one focus.

In the seventeenth century, René Descartes devised a means of plotting conic (and other) curves on graph paper. This did wonderful things for mathematics, but it removed conic curves from the realm of pure geometry and placed them in the realm of mathematics. At present, nearly everyone thinks of these curves in terms of columns of numbers and dots on graph paper. We are sometimes amazed to find that they all occur in nature. The planets move freely in elliptical orbits without once thinking that they must move over so many x units and down so many y units. They simply move freely in balance with the forces influencing their motion. The same applies to the hyperbola described by a high-speed comet making a single pass by the sun or the parabola described by a nearsighted monkey that misses the branch for which it was jumping. The natural occurrences of conic curves are numerous. Lamps cast hyperbolic patterns against living room walls, and a soccer ball casts an elliptical shadow as it is thrown in the early morning sunlight. The frequency at which these curves occur naturally and the diverse way of generating them may substantiate the belief of the ancients that conic curves were of divine origin.

To reduce these curves to formulas plotted on graph paper is like reducing a work of art to a "paint-by-number" scheme. It is true that a fine painting can be reduced to a computer readout that gives point positions and color numbers, allowing it to be reconstructed in great detail as a color television picture, but it would be as wrong to identify a painting with a computer readout as it is to identify a conic curve with a formula.

The traditional tools of geometric construction are the compass and the straightedge. To these we shall add the square. Most of the constructions shown in this book involve right angles; although it is possible to construct perpendiculars in each case, it is much more convenient to use a square. The transparent triangles used in drafting are excellent for this purpose. A 45°-45°-90° triangle can be made even more useful by scribing a line from the right angle to the midpoint of the hypotenuse. All the constructions shown in this book can easily be drawn by students with only a compass, a right triangle or square, and a very sharp pencil. For accuracy, a number 3 pencil is excellent, but it should be sharpened frequently.

FIGURE 0-3

Drawing tools.

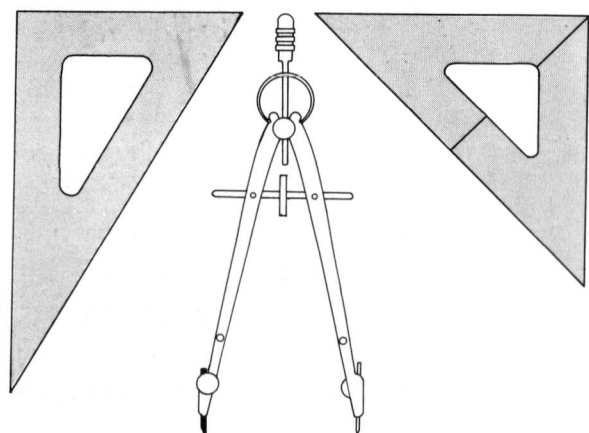

It should not be surprising that practical applications of conic curves were not pursued at the time that they were studied the most intensely. Ancient people felt that abstract knowledge (philosophy) was the only thing that mattered. The practicalities of inventions based on their theoretical conclusions were not just overlooked but were actively looked down on. This attitude is stated rather bluntly by Lucius Annaeus Seneca (4 B.C.–A.D. 65), who felt that inventors had no class at all:

> In my own time there have been inventions of this sort, transparent windows, tubes for diffusing warmth equally through all parts of a building, short-hand, which has been carried to such a pitch of perfection that a writer can keep pace with the most rapid speaker. But inventing of such things is drudgery for the lowest slaves; philosophy lies deeper....

This attitude generally prevailed until the industrial revolution joined logic and practicality, but vestiges of it still surface in some our educational institutions.

This book has been written to give the practical aspects of the construction and uses of conic curves. I have tried to avoid the tedious proofs that abound in most books on the subject, including the *Encyclopaedia Britannica,* where we can find Pascal's theorem for hexagons inscribed in conic curves and how to draw a conic through any five points, no three of which lie on the same straight line. The ultimate in unintelligibility was written in the seventeenth century by Sir Thomas Browne in *The Garden of Cyrus:*

> For spherical bodies move by fives, and every globular figure placed upon a plane, in direct volutation, returns to the first point of contaction in the fifth touch, accounting by the Axes of the Diameters or Cardinall points of the four quarters

thereof. And before it arriveth unto the same point again, it maketh five circles equall unto itself, in each progress from these quarters absolving an equal circle.

Those who wish to pursue the impractical aspects of conic curves may refer to the bibliography.

CHAPTER ONE

DERIVING ELLIPSES

METHOD 1

At the risk of being obvious, ellipses (and the other conic sections) may be obtained by cutting up (sectioning) a cone. Although this may not be the most convenient way of obtaining an ellipse, it must be listed as a legitimate means of deriving one. The intersection of a cone and a plane that passes completely through the cone is an ellipse. Ellipses are also generated at the intersection of a cylinder and a plane, but a cylinder must be considered to be a part of a special kind of cone having an apex angle of 0°. Figure 1-1 shows the shadow of a ball illuminated by a point source of light. The shadow cast on the table is an ellipse, with the ball touching the surface at one focus. (The shadow of a sphere is always conical, regardless of the angle from which the sphere is illuminated.)

FIGURE 1-1

An elliptical shadow cast by a ball. The ball rests on one focus of the ellipse (proof given in Chapter 9).

METHOD 2

Ellipses occur naturally in free orbital motion. Such motion ranges from planets having nearly circular orbits to the extremely eccentric orbits of recurrent comets.

METHOD 3

For those who enjoy working algebra problems and putting dots on graph paper, the equation $\frac{x^2}{a^2} + \frac{y^2}{b^2} = 1$ describes an ellipse in the *xy* plane with major and minor axes of length $2a$ and $2b$. The standard nomenclature for an ellipse described in analytical geometry is shown in Figure 1-2.

FIGURE 1-2

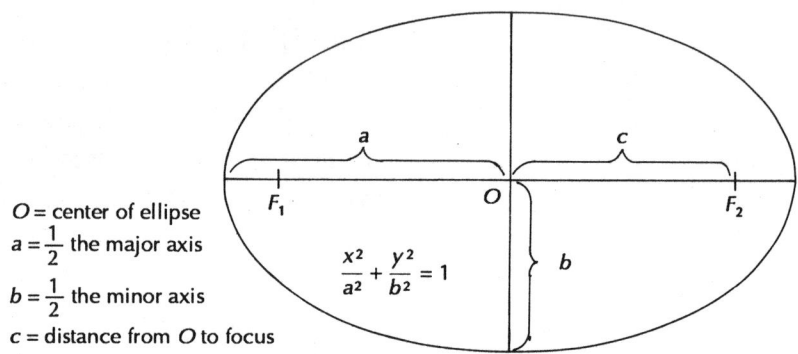

O = center of ellipse
$a = \frac{1}{2}$ the major axis
$b = \frac{1}{2}$ the minor axis
c = distance from O to focus

$$\frac{x^2}{a^2} + \frac{y^2}{b^2} = 1$$

METHOD 4

Ellipses may be defined as the locus of a point, the sum of whose distances to two fixed points is a constant. Put into practice, this method resolves itself to the two-pins-and-a-string method of constructing ellipses. Two pins are placed at the foci and a loop of string is adjusted to a length that allows the pencil point to touch a point on the ellipse. This point is usually at the major or minor axis intercept, but it may be any point known to be on the ellipse. See Figure 1-3.

FIGURE 1-3

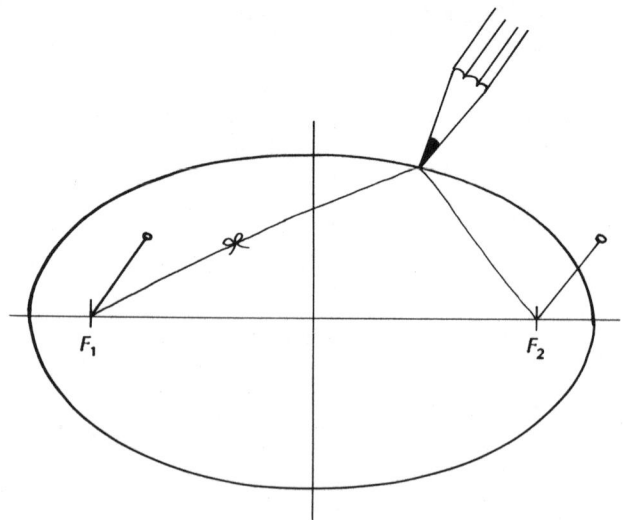

2 | PRACTICAL CONIC SECTIONS

This is a very practical way of drawing ellipses, and it is often the most convenient approach to be used for laying out large ellipses, such as elliptical flower beds or large outdoor signs.

It is possible to accomplish the same thing without the use of pins and string. Going back to the definition of an ellipse as the locus of a point whose distance to two fixed points is a constant, we should establish the fact that this constant is always equal to the length of the major axis of the ellipse. If we establish the major axis on a line and mark off arbitrary points along this line, we may take the distances (with a compass) from point A to one of these points and swing an arc from F_1, as shown in Figure 1-4. From point B we adjust the compass to measure the length from B to the same point and swing another arc from F_2. The intersection of the two arcs will be a point on the ellipse. By repeating this operation several times and connecting these points of intersection, we may draw the ellipse. Although this appears to be a practical way to draw an ellipse, in practice it becomes difficult to draw through the points as we approach the ends of the ellipse.

FIGURE 1-4

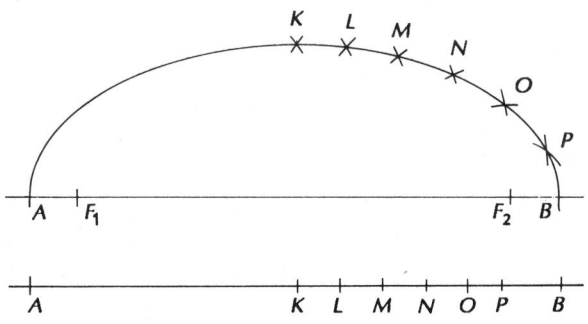

It is important to remember that the constant is always equal to the length of the major axis of the ellipse. In Chapter 2 we shall see that the constant involved in generating hyperbolas is also equal to the major axis (the distance between vertices) of a pair of hyperbolas, the only difference being that we subtract the two distances instead of adding them when determining points.

CHAPTER ONE: DERIVING ELLIPSES

METHOD 5

The *trammel method* is an easy way to draw ellipses; it requires no pins or construction lines except the major and minor axes. For this reason it is frequently preferred by drafters. Two approaches may be used. In Figure 1-5(a) one-half the lengths of the major and minor axes are marked off on a piece of cardboard or plastic and placed over two lines drawn perpendicular to each other. The point P will be on the ellipse as long as the points M and N are on the x and y axes. Similarly, Figure 1-5(b) shows a trammel marked with one-half the minor axis inside one-half the major axis. Again, if the points M and N are positioned over the axis lines, point P will fall on the ellipse.

FIGURE 1-5

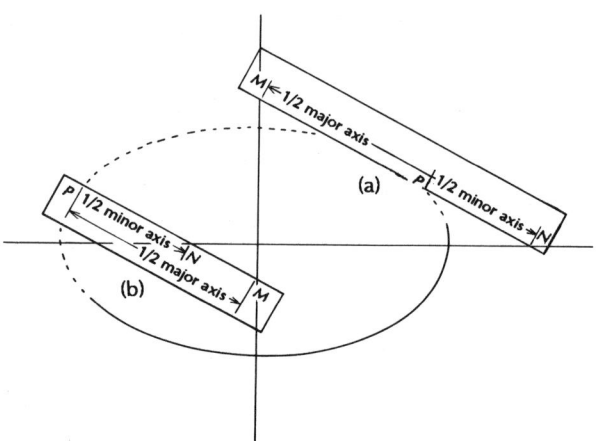

A mechanical device known as an ellipsograph, or the trammel of Archimedes, is used for drawing ellipses and is shown in Figure 1-6. The pen (P) is shown at the end of the movable arm, but any point on the arm will describe an ellipse. Note that this method is no different from the method shown in Figure 1-5(b) but is presented in a more practical mechanical form. The point P will cross the major axis when M is centered and will cross the minor axis when N is centered.

FIGURE 1-6

Ellipsograph.

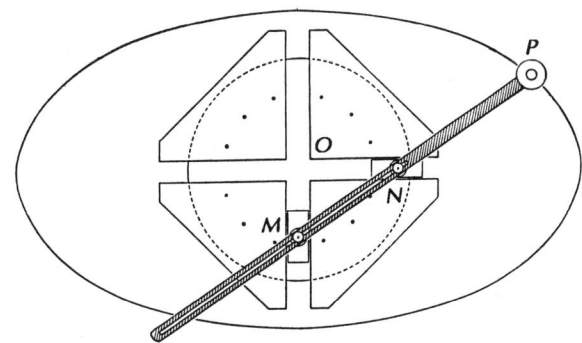

METHOD 6

The parallelogram method starts with a pair of intersecting axes centered on a parallelogram that is to circumscribe the ellipse. Divide AO and AE into the same number of equal parts, as shown in Figure 1-7. From D, draw lines through points 1, 2, and 3 on AO; and from C, draw lines through points 1, 2, and 3 on AE. The intersections of these lines will be points on the ellipse. Although any parallelogram will work, it is more convenient if the parallelogram is a rectangle; otherwise the axes will not correspond to the major and minor axes of the ellipse.

FIGURE 1-7

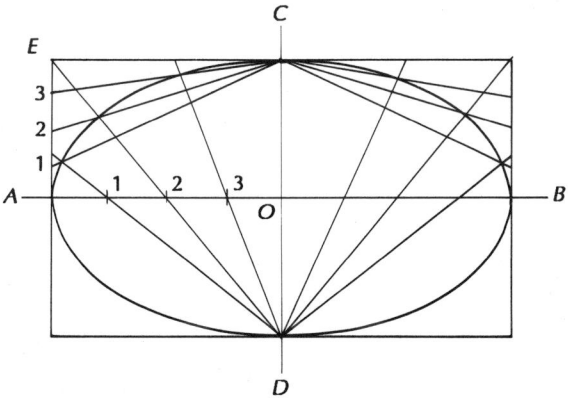

METHOD 7

The *Directing Circle* method has several advantages over the other methods described. It gives tangent lines instead of points to connect and has the further advantage of being part of a system for drawing all conic curves. This method of deriving conic curves is so important that an entire chapter (Chapter 4) is devoted to it.

Draw a circle with a diameter equal to the major axis of the desired ellipse. Draw the diameter line that will also be the major axis of the ellipse; then establish the foci at two points equidistant from the center. Place a drafting triangle over the circle so that one edge passes through one focus and the right angle falls on some point of the circle. The other arm of the right angle will be tangent to an ellipse at some point. By drawing several tangent lines, an ellipse will be derived inside the directing circle (Figure 1-8). Although either focus may be used, it is best to use the one more distant from the tangent, since the near focus is so close to the tangent that it is difficult to maintain accuracy.

FIGURE 1-8

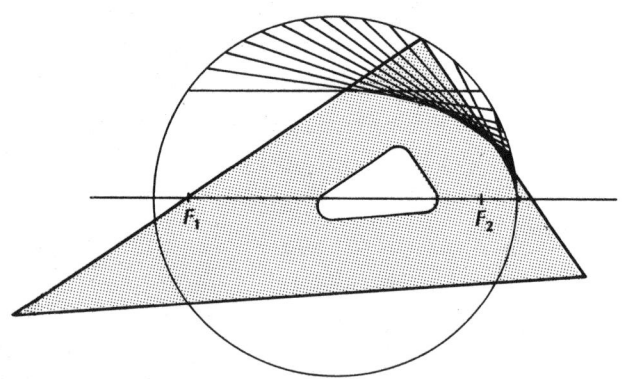

Another advantage of this method is that it allows us to derive unstated parameters quickly and easily. For example, if the foci and major axis are known and we wish to find the length of the minor axis, we simply construct a perpendicular at either focus. The length of the line F_1P_1 (Figure 1-9) is identical to the minor axis. This follows logically, since a perpendicular to this line constructed at P_1 will be the tangent to the ellipse that is parallel to the major axis, which can only be at the intercept of the minor axis.

FIGURE 1-9

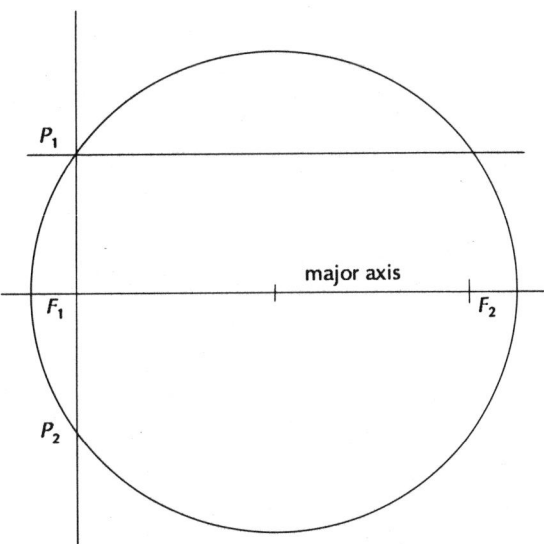

If only the major and minor axes of an ellipse are given, the foci may be found by swinging an arc whose radius is equal to one-half the major axis (the distance a) from either minor-axis intercept. The two intersections of the arc and the major axis will be the foci. Many people do not realize that this is exactly what they have been doing mathematically by establishing foci with the formula $c^2 = a^2 - b^2$. The resulting right triangle has a hypotenuse equal in length to one-half the major axis (a), the distance b is given, and the distance c must satisfy the Pythagorean theorem (Figure 1-10). The advantage of establishing foci with a compass is that numerical values need not be assigned to any of these dimensions.

FIGURE 1-10

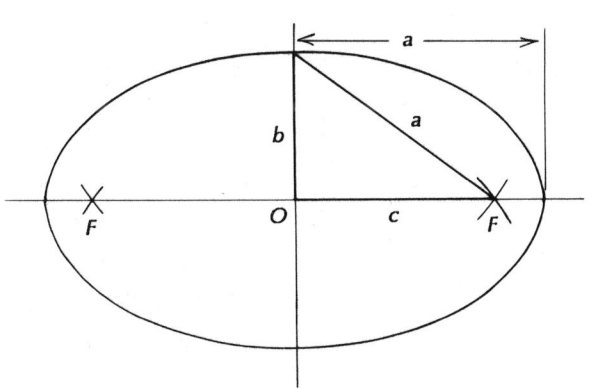

CHAPTER ONE: DERIVING ELLIPSES

METHOD 8

The two-circle method may also be useful to drafters although it requires the use of construction lines. Draw two concentric circles, the smaller having a diameter equal to the minor axis and the larger having a diameter equal to the major axis. Draw several radials extending through both circles. Next, draw a horizontal line at the point of intersection of each radial and the smaller circle and a vertical line at its intersection with the larger circle. The intersection of the two perpendicular lines drawn from each radial is a point on the ellipse. See Figure 1-11.

FIGURE 1-11

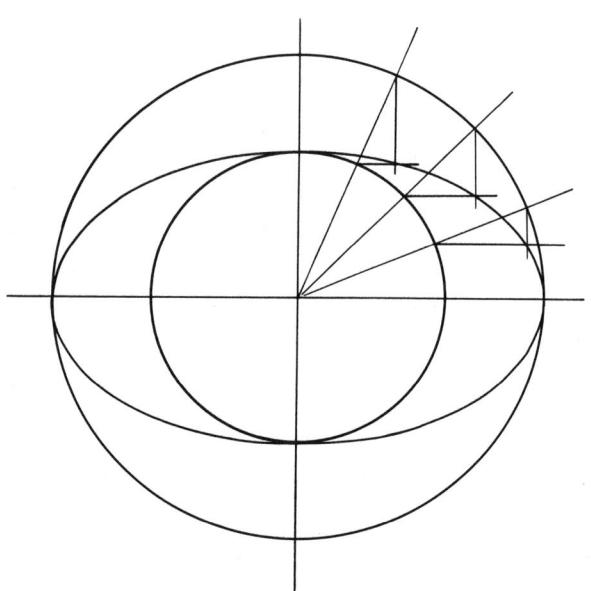

METHOD 9

Ellipses show up in another way that is not really practical for construction but merits some discussion. Two sources of harmonic motion (sine waves) that are of the same frequency but are not in phase with each other may be described as having elliptical properties. An oscilloscope fed by two such wave sources (one to the vertical axis and the other to the horizontal axis) will display an ellipse. If the two signals have exactly the same amplitude and have a phase difference of 90° or 270°, the figure, called a *Lisajous figure,* will be a circle—but remember that circles are special ellipses having equal major and minor axes. If the phase angles or the amplitudes are changed, an ellipse will be traced on the oscilloscope.

In optics and microwave transmission, when two components of transmitted energy are displaced in phase and in space, the signal is described as having elliptical or circular polarization.

CHAPTER TWO

DERIVING HYPERBOLAS

METHOD 1

A plane cutting a cone that does not pass completely through the cone no matter how far the cone and the plane are extended will produce a curve that increases in width to infinity. Since the curve will never close, there is no formula for its area. When such a curve is formed by a plane that does not cut the cone parallel to its side, the curve is a hyperbola. Although the plane never leaves the cone in one direction, it will intersect the other nappe of the cone to form a pair of hyperbolas. Whether the other nappe of the cone is real or imagined, this second curve must be considered. Since construction of a pair of solid cones joined at the apex is difficult, a lamp can be used to illustrate hyperbolas.

FIGURE 2-1

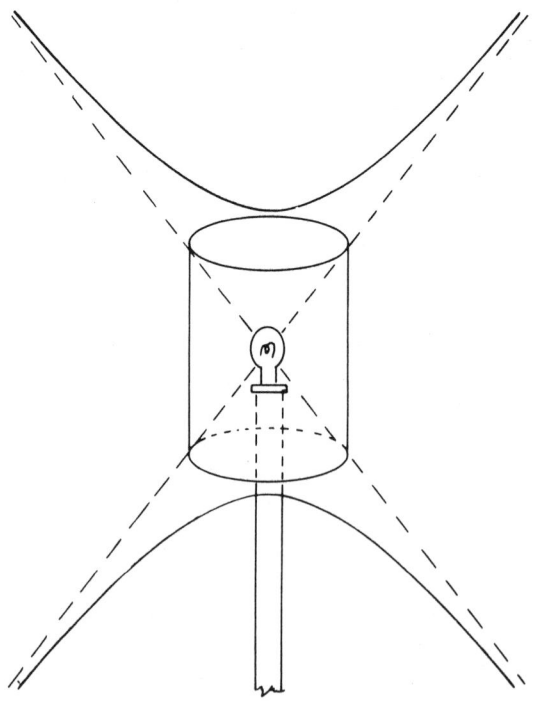

Figure 2-1 shows an ideal situation in which the lampshade is cylindrical and the light bulb is located precisely in the center of the shade. In this case, a pair of identical hyperbolas is cast on the wall. Usually, however, the shade is conical and the bulb is located arbitrarily. The resulting hyperbolas will be accurate but will not be geometrically or mathematically related. Should the lamp be tilted toward the wall in such a manner that the far (from the wall) edge of the cone is parallel to the wall, the curve cast on the wall will be a parabola. If the lamp is tilted farther, an ellipse will be shown.

METHOD 2

Mathematically, Cartesian coordinates for hyperbolas may be plotted using the formula $\frac{x^2}{a^2} - \frac{y^2}{b^2} = 1$. (A hyperbola whose asymptotes are the xy axes may be drawn by plotting $xy = k$, where k is any constant.) The standard nomenclature used in analytical geometry for a hyperbola is shown in Figure 2-2.

FIGURE 2-2

O = center
$a = \frac{1}{2}$ the major axis
$b = \frac{1}{2}$ the minor axis
c = distance from O to focus

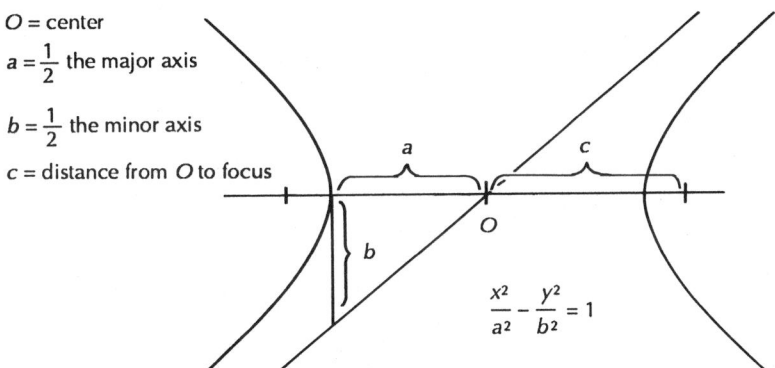

METHOD 3

Hyperbolas are infrequently observed in nature. Comets having orbital eccentricities greater than unity follow a hyperbolic path. Since hyperbolas are open-ended figures, these comets will not return.

METHOD 4

The interference of two systems of radiating concentric circles forms a family of hyperbolas, as shown in Figure 2-3. This is best illustrated by dropping two stones into a quiet pond and watching the ripples as they cross each other. The interference of two sources of high-frequency energy will generate a hyperbolic pattern of reinforcement and cancellation. In *Optics of the Electromagnetic Spectrum,* Dr. Charles Andrews describes tracing a family of hyperbolas on a school lawn using a meter to detect the lines of minimum intensity of two interfering radio signals (of the same frequency) and marking these with a tennis-court line marker.

FIGURE 2-3

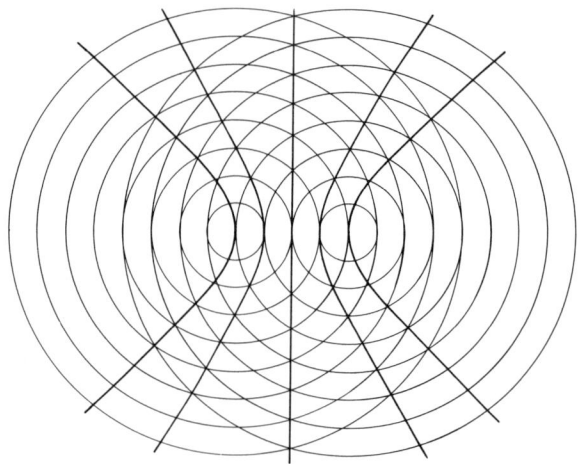

METHOD 5

A hyperbola may be defined as the locus of a point, the difference of whose distances from two fixed points (the foci) is a constant. As might be expected, a device may be constructed somewhat similar to that described in the two-pins-and-a-string method (Chapter 1) for constructing ellipses. Since this method, illustrated in Figure 2-4, is considerably more complicated and requires drilling holes in the drawing table, it is not likely to be the most popular way to draw hyperbolas. The crank mounted below the table extends or retracts the same length of string through each hole. The difference between m and n remains constant, thus complying with the definition of a hyperbola.

FIGURE 2-4

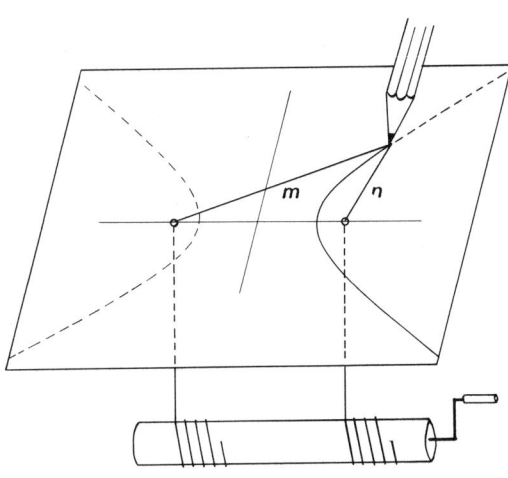

METHOD 6

There is another method of constructing a hyperbola that is similar to Method 5 but does not require drilling holes in the drawing board. It also utilizes the principle that a hyperbola is the locus of a point, the difference of whose distances from two fixed points is a constant.

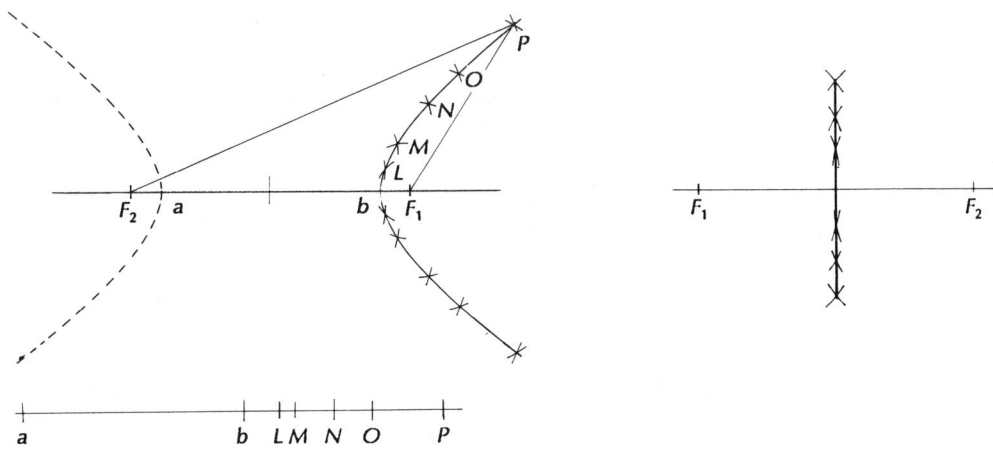

FIGURES 2-5 AND 2-6

CHAPTER TWO: DERIVING HYPERBOLAS

Once we have decided on the distance between foci and the distance ab, which serves both as a distance between vertices and as the constant, we may construct the hyperbola. Draw a line slightly longer than the distance between foci. Mark a and b (the vertices) and F_1 and F_2 on this line. The center point is not used except for laying out these points symmetrically. As an aid to construction, draw another line (apart from the drawing), mark the distance AB, and add the arbitrary points $L, M, N, O,$ and P. (At this point it would be helpful to have two compasses, since the adjustment from F_1 to F_2 is considerable.) Place the compass at the point marked B and measure the distance BL. Transfer this to the drawing, placing the compass point at F_1 and drawing an arc. Again place the compass on the auxiliary line at the point marked A and measure the distance AL. Place the compass point at F_2 on the drawing and draw an arc that intersects the arc drawn from F_1. This will be a point on the hyperbola. In a similar manner, repeat the process from points $M, N, O,$ and P. Connecting these points will give a hyperbola. See Figure 2-5.

If we set the distance between vertices of a pair of hyperbolas at zero, the resulting "curve" will be a straight line perpendicular to the axis, as shown in Figure 2-6. This follows all of the geometric rules for a hyperbola and has been used in compound reflective antennas. We will discuss this further in Chapters 5-6.

METHOD 7

The parallelogram method of constructing hyperbolas is similar to that shown for ellipses in the last chapter. If we know the distance AB and one point D on the hyperbola, we may construct the rectangle $DEGF$. Divide lines DC and DF into the same number of equal parts and number them from point D. From point A, draw lines to points 1, 2, and 3 on DC; and from point B, draw lines to points 1, 2, and 3 on DF. The points of intersection of similarly numbered lines may be connected to form a hyperbola (Figure 2-7). This method will work when the construction lines are not perpendicular, but the lines will not be axes of the hyperbola, so it is more practical to use a rectangle instead of another parallelogram.

FIGURE 2-7

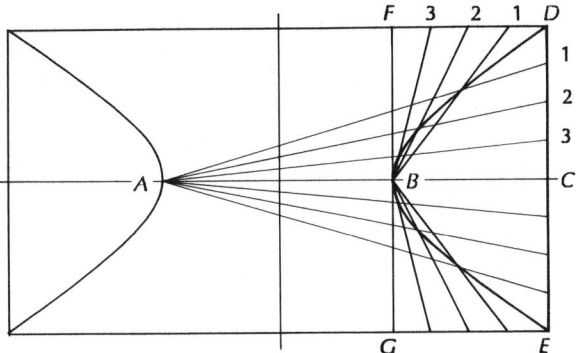

METHOD 8

The Directing Circle method is practical and has advantages over most of the other methods of drawing hyperbolas. As discussed in Chapter 1, this method uses tangent lines instead of points.

Draw a circle that will touch the hyperbola at the vertices. Draw a line through the circle, extending beyond the circle in both directions. Using the center of the circle, select the focal length and mark the foci with a compass. Place a right angle over the circle positioned so that one arm of the right angle passes through one focus and the right angle occurs on the directing circle. The other arm of the right angle will be tangent to the hyperbola. Again, as with the ellipse, it is more convenient to use the focus that is farther from the hyperbola being drawn, but the other focus may be used. When a representative number of tangents have been drawn, the curve envelope takes shape, as shown in Figure 2-8.

The asymptotes of a pair of hyperbolas may be found by placing the right angle over the circle so that one arm passes through a focus and the other arm passes through the center of the circle. This is discussed in greater detail in Chapter 4.

FIGURE 2-8

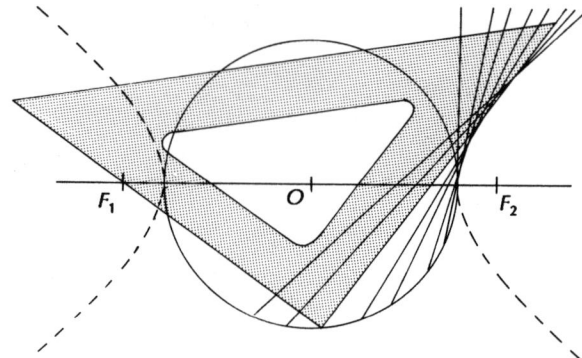

CHAPTER THREE

DERIVING PARABOLAS

METHOD 1

A parabola is described when a plane intersects a cone parallel to one side of the cone. The plane never passes completely through the cone, as in the case of the ellipse, nor does it intersect the other nappe of the cone, as it does in the case of the hyperbola. This places parabolas between ellipses and hyperbolas in a manner analogous to noon dividing the morning and afternoon. This means that there are really only two kinds of conic curves—open and closed. Although the circle is a special case of the ellipse, the parabola is a special case of both the ellipse and the hyperbola. As we might expect, parabolas have the geometric and reflective properties of ellipses and hyperbolas and may be thought of in these terms if we consider the second focus to be infinitely distant.

METHOD 2

Parabolas occur naturally when an object traveling at a constant velocity is acted upon by gravity. For practical purposes, this is true for objects propelled a short distance, such as a rock thrown from a window or a stream of water from a hose. We are assuming that Earth's center of gravity (at some 4,000 miles) might as well be at infinity.

FIGURE 3-1

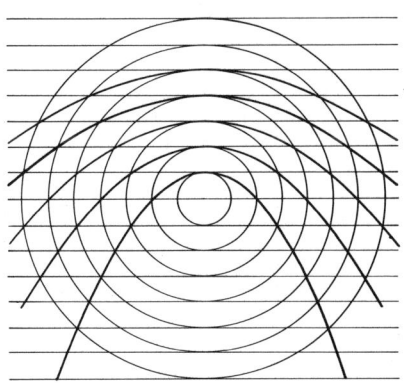

METHOD 3

The pattern created by the interference of circular waves and parallel waves is a group of parabolas as shown in Figure 3-1. This is similar to the family of hyperbolas formed by the interference of two circular wave systems shown in the section on hyperbolas. The difference is that the center of the parallel waves is infinitely distant.

METHOD 4

As we might expect, there is a mechanical method of drawing a parabola similar to the methods shown for an ellipse and a hyperbola. Draw an axis line *AB* and a perpendicular line *CD* approximately half the distance between the points *F* and *B*. Place a pin at *F* and attach a string from this point to *B* and back to *O*. Using a drafting triangle to maintain perpendicularity to line *CD*, move the string from point *O* toward *C* and *D*. The point *P* will describe a parabola having a focus at *F* (Figure 3-2). Although this method will yield a parabola, it is not really one of the more practical ways of doing so. This method may find use in checking the accuracy of paraboloids (parabolas of rotations), since all points and lines are located on one side of the paraboloid.

FIGURE 3-2

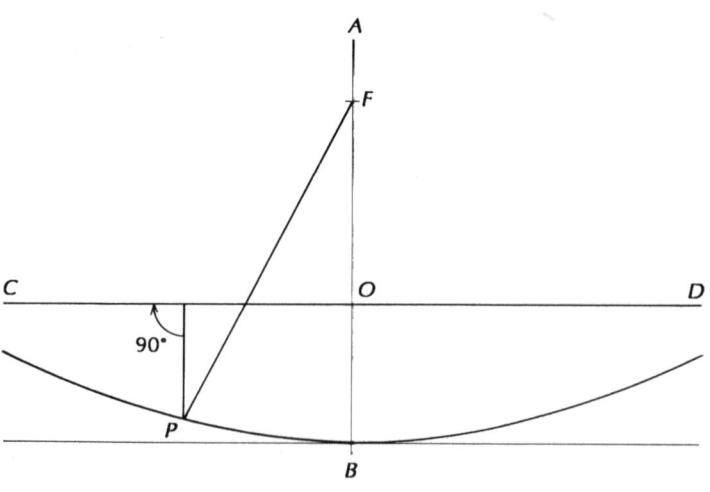

METHOD 5

A parabola may be formed by connecting points on two intersecting lines. Draw two lines at any convenient angle between 0° and 180°. Mark off points at any constant interval, starting from the point of intersection. Number these points in opposite directions starting at point *A* and numbering toward *C* and starting at *B* and numbering toward *A* (Figure 3-3).

FIGURE 3-3

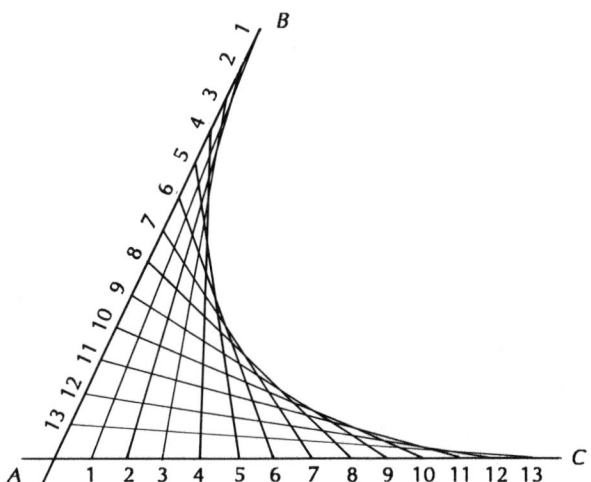

Connect similarly numbered points with lines, and the ensuing envelope of tangent lines will describe a parabola.

Lines *AB* and *AC* need not be of equal length as long as they are divided into the same number of parts. However, the bisector of ∠*BAC* will be the axis of the parabola only if lines *AB* and *AC* are equal.

These figures are frequently constructed by students using pins and colored thread. Although they are true parabolas, the curves are hard to predict and the focus is not defined. This method has few practical applications.

METHOD 6

The parallelogram method may be used if we know the axis

AB and a point *C* on the parabola. Draw a rectangle *CDFE* and divide *CB* and *CE* into the same number of equal parts, numbered from point *C*. Connect point *A* to points 1, 2, and 3 on line *CE*, and draw lines parallel to *AB* through points 1, 2, and 3 on line *CB*. The intersection of similarly numbered lines may be connected to form a parabola (Figure 3-4). Again, any parallelogram may be used for this construction, but the line *AB* will not be the axis of the parabola if *AB* is not perpendicular to *CD*.

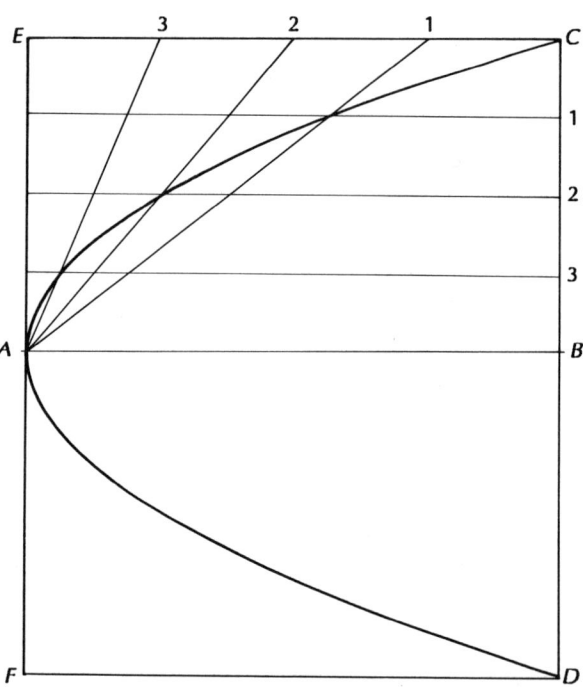

FIGURE 3-4

METHOD 7

A parabola may be defined as the locus of a point whose distance from a fixed line is equal to its distance from a fixed point. The fixed line is called the *directrix* and the point is the focus. The directrix is perpendicular to the axis of the parabola and is located exactly the same distance on one side of the vertex as the focus is on the other. Figure 3-5 shows a parabola with focus, directrix, and point *P*, which moves in a path equidistant from the focus and the directrix.

FIGURE 3-5

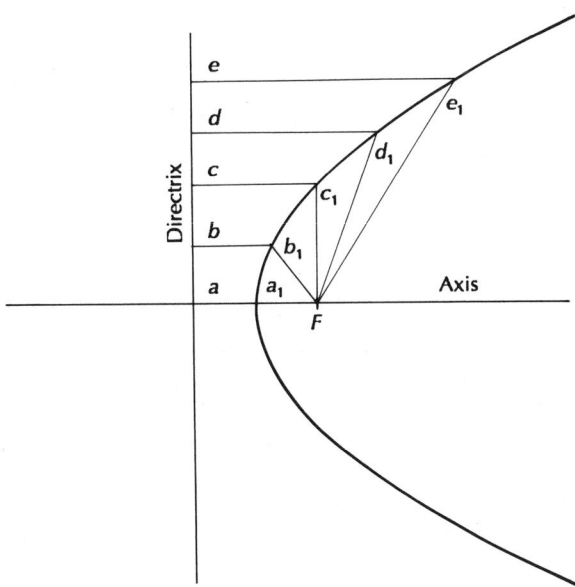

Although it would appear at first that there must be a simple way to construct a parabola geometrically (similar to the two-pins-and-a-string method for ellipses) using directrix, it really is not that simple, since the parabola is an open figure.

METHOD 8

Mathematically speaking, the coordinates for a parabola may be plotted using the formula $y^2 = 4fx$. Actually, any quadratic equation will describe a parabola when graphed. The standard nomenclature used in analytical geometry for a parabola is shown in Figure 3-6.

FIGURE 3-6

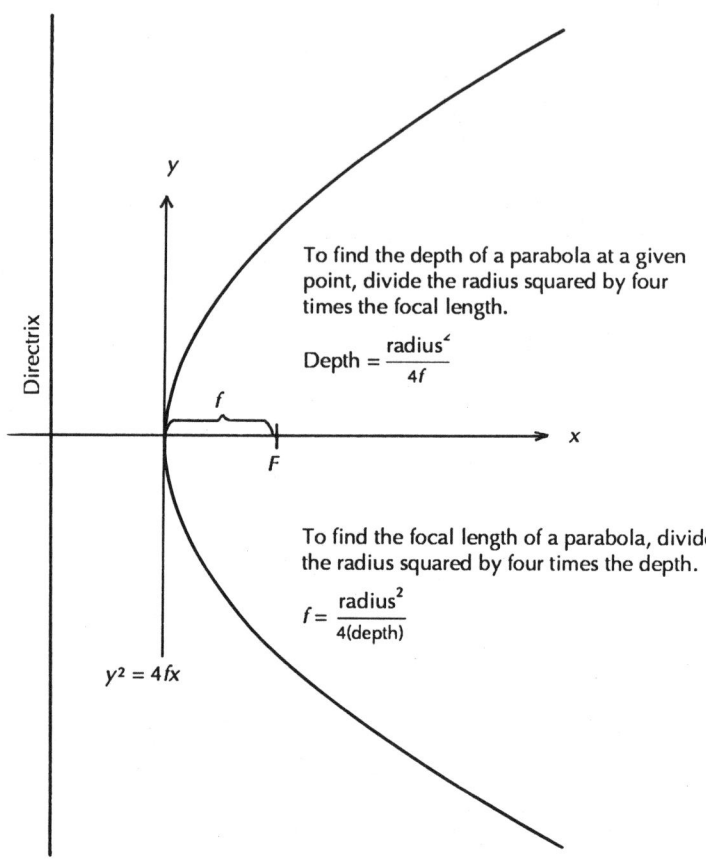

METHOD 9

A paraboloid (a parabola of rotation) may be formed by carefully spinning a shallow container of liquid. Although this seems to be an awkward way of making a paraboloid, it has been used successfully. At least one astronomical telescope has been made by slowly spinning a pan of mercury to form a parabolic reflector. The obvious disadvantage of this reflector is that it must always look straight up, and a series of optically flat reflectors must be positioned to permit the viewing of other areas of the sky.

Scientists at the Steward Observatory of the University of Arizona have embarked on an ambitious project to spin-cast large glass mirrors for astronomical telescopes. These are of cellular (honeycomb) design to reduce weight and are much deeper than any single-piece mirrors ever built. The engineering problems in such a project are astronomical. To cast an 8-meter (26-foot) reflector, the glass is melted in a mold inside an 1180°C (2156°F) oven for 40 hours. During this period, both the mold and the oven rotate at around 7 revolutions per minute. After a 2- to 3-month annealing and cooling cycle, the mirror blank is removed from the oven and is ready for polishing. Because it is spin-cast, the surface of the mirror blank is already a paraboloid. This saves the time and expense of rough grinding away as much as 28 metric tons of glass, had the mirror been cast with the traditional flat surface. The entire process of spin-casting and fine polishing takes less than two years.

METHOD 10

The Directing Circle method described in the chapters on ellipses and hyperbolas applies also to parabolas. The only difference in the case of the parabola is that the "circle" must be infinitely large and is therefore represented by a straight line. This may come as some relief to those who pondered what happened to the conic curve having an eccentricity of 1.00 in the previous two examples of the Directing Circle method of generating conic curves. As the foci approached the Directing Circle, the ellipse became flatter, and it was obvious that at the circle, the ellipse would have only a major axis. This problem no longer exists if the diameter of the Directing Circle is infinite. We may now place the focus anywhere on the axis and draw a parabola. Perhaps it is better to say "the parabola," since all parabolic curves are the same—only the size changes. This may be said of any conic curves having the same eccentricity. For example, all ellipses having an eccentricity of 0.85 are the same shape, although their sizes may vary. Since all parabolas have an eccentricity of 1.00, the only difference can be in their size.

A parabola may be derived by drawing two perpendicular lines, one to be the axis and the other the tangent to the vertex of the parabola. A point may be placed anywhere on the axis. This will be the focus. Place a triangle over the paper so that one edge of the right angle passes through the focus and the right angle occurs on the perpendicular line. The other edge of the right angle will be tangent to the parabola. The point of tangency will be exactly twice the distance of the right angle to the axis (Figure 3-7). This method of generating parabolas is certainly as accurate as working out a system of coordinates and connecting points.

FIGURE 3-7

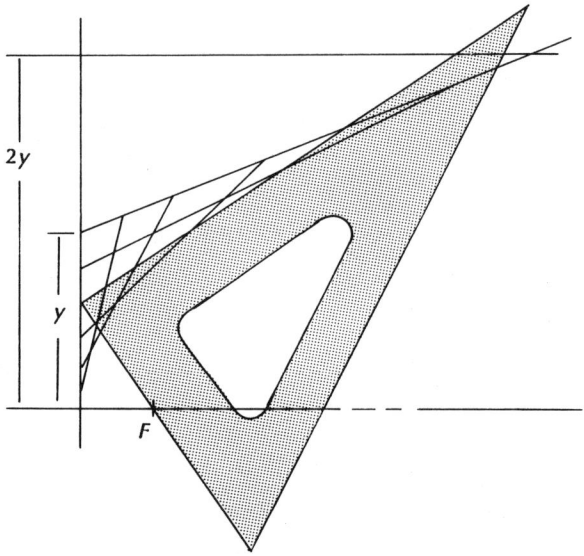

24 | PRACTICAL CONIC SECTIONS

CHAPTER FOUR

THE DIRECTING CIRCLE

The Directing Circle method of constructing conic curves is by far the most practical of all of the construction methods shown in the previous chapters and is the best method for developing a visual concept of these curves. It seems strange that it is only rarely presented in geometry books and then usually is used to prove a mathematical premise rather than to demonstrate a universal method for deriving all conic curves. Mathematicians have dismembered this logical family of curves into four formulas:

$x^2 + y^2 = c^2$	$\dfrac{x^2}{a^2} + \dfrac{y^2}{b^2} = 1$	$y^2 = 4fx$	$\dfrac{x^2}{a^2} - \dfrac{y^2}{b^2} = 1$
Circle	Ellipse	Parabola	Hyperbola

It is doubtful that a student can visualize a family of curves from these seemingly unrelated formulas. The only apparent relationship found in them is the change of + to − in the equations describing ellipses and hyperbolas. It certainly is not obvious that the circle is a special case of an ellipse and that the parabola is a special case of both the ellipse and the hyperbola. Actually, a few minutes spent shining a flashlight on a wall at various angles can give you a better understanding of the continuity of the family of conic curves than an entire book of formulas.

To develop an understanding of these curves, it is recommended that you take a compass, a drafting triangle, a sharp pencil (3H or harder), and a stack of blank paper and spend several hours drawing various curves using the Directing Circle method. There is no need to measure anything. Simply choose arbitrary positions for the foci and the radius of the Directing Circle and see what happens. Use your compass to make sure that the foci used in ellipses and hyperbolas are located symmetrically. In the interest of conservation of paper, it is possible to use the same circle to draw an ellipse (inside) and a hyperbola (outside).

After constructing a number of curves by this method, some interesting relationships emerge. Although it is difficult to predict the exact point of tangency (except for parabolas) using this method of construction, it is possible to pinpoint them in three places on ellipses (at major and minor axes and at the focal width) and two places on hyperbolas (the vertex and focal width). When we find the tangent of a conic at its focal width, we find that the slope of this tangent line is equal to the eccentricity of the curve. The focal width (*latus rectum*) is the distance across the curve at the focus (Figure 4-1). For parabolas, the focal width is always four times the focal length. For ellipses, it is less than four, and for hyperbolas, it is more than four.

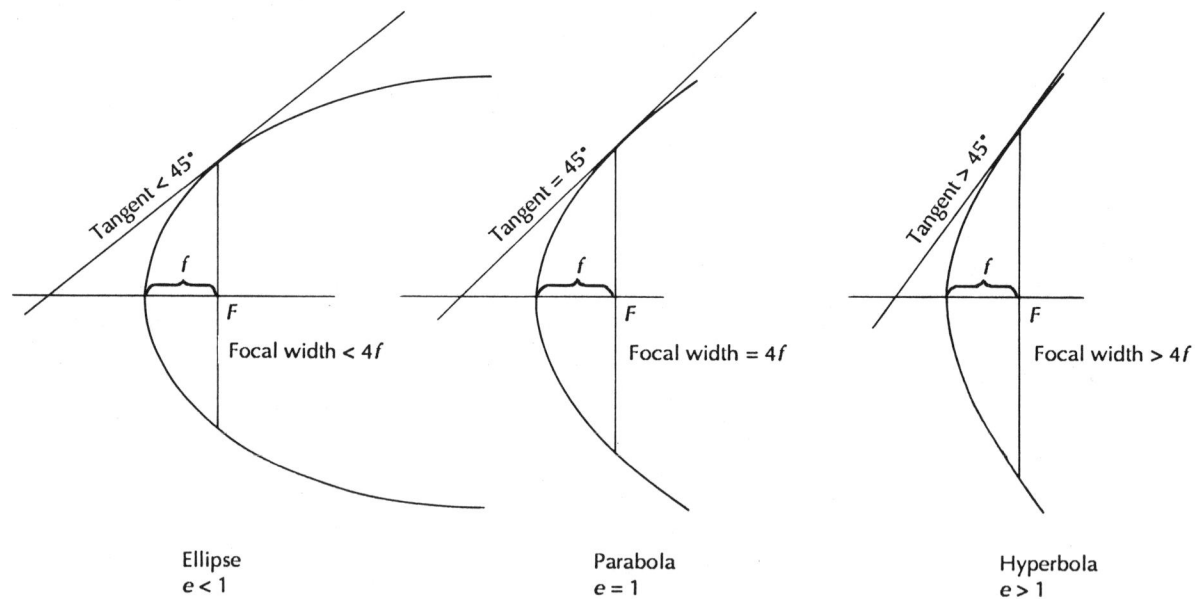

FIGURE 4-1

Tangents at focal width.

CONSTRUCTING AN ELLIPSE

Using the Directing Circle method, let us first construct an ellipse. After deciding on a radius of the circle (A) and the location of the foci (C), we position our drafting triangle so that one arm of the right angle is directly over F_2 and the right angle is directly over some point on the circumference of the circle. The other arm of the right angle will, at some point, be tangent to the surface of an ellipse. We draw enough tangent lines to see the shape of one quadrant of the ellipse. Remember that if we have one quadrant of an ellipse, we have everything that we really need to know about the ellipse. See Figure 4-2.

FIGURE 4-2

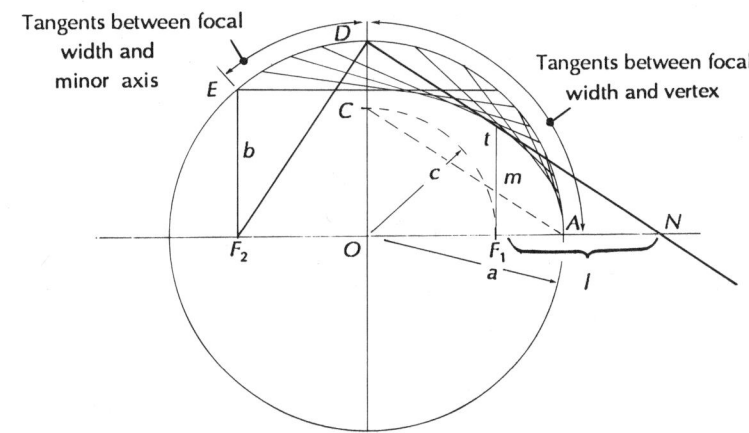

When we place the right angle over the Directing Circle so that one arm passes through F_2 perpendicular to the major axis, the right angle will pass through the circle at point E. The line F_2E is equal to the length of the minor axis of the ellipse, but it is displaced from its proper place by the distance c. The logic of this is obvious when we consider that the tangent to the ellipse at its minor axis is parallel to the major axis and that the angles at F_2 and E are right angles. It is also obvious that all possible tangents for the quadrant of this ellipse will occur between E and the intercept of the major axis at the right end of the ellipse.

At this point we should draw the minor axis perpendicular to O and extend it through the Directing Circle at point D. When we place our right angle over the circle so that one arm passes through F_2 and the right angle is over D, the other arm of the right angle will be tangent to the curve at its focal width. We extend this line to point N on the extended major axis of the ellipse. All tangents drawn in the quadrant of the circle from D to the major-axis intercept will lie on the curve from its focal width to its vertex. All tangents from the focal width to the minor-axis intercept will be generated by placing the right angle over the arc from D to E.

To prove that the slope of the ellipse at its focal width is equal to its eccentricity $e = \frac{c}{a}$, it is necessary only to prove two right triangles to be similar.

We construct a line parallel to DN passing through the point A, which is the vertex of the ellipse as in Figure 4-2. The point at which this line intersects the minor axis (or in some cases the projection of the minor axis) will be found by compass to be equal to the distance c. Since the newly constructed line is parallel to the tangent at the focal width and two sides of the newly formed triangle are equal to c and a, we have $e = \frac{c}{a}$, which is both the eccentricity of the ellipse and the slope of the tangent line at its focal width. Chapter 7 contains a detailed description of eccentricity.

CONSTRUCTING A HYPERBOLA

The same relationship holds true for hyperbolas (Figures 4-3 and 4-4). In a manner similar to the last example, let us construct a hyperbola. When we place the right angle over the circle with one arm passing through F_2 and the right angle over point D at the top of the Directing Circle, we find that the other arm of the right angle becomes the tangent to the hyperbola at its focal width. All the tangents generated by placing the right angle over the arc between D and E will lie between the focal width and the asymptote. (As noted in Chapter 2, an asymptote is defined by positioning the right angle over the Directing Circle so that one arm passes through a focus and the other passes through the center of the circle while the right angle is positioned over the Directing Circle.) All the tangents generated by placing the right angle over the arc between *D* and *A* will lie between the focal width and the vertex.

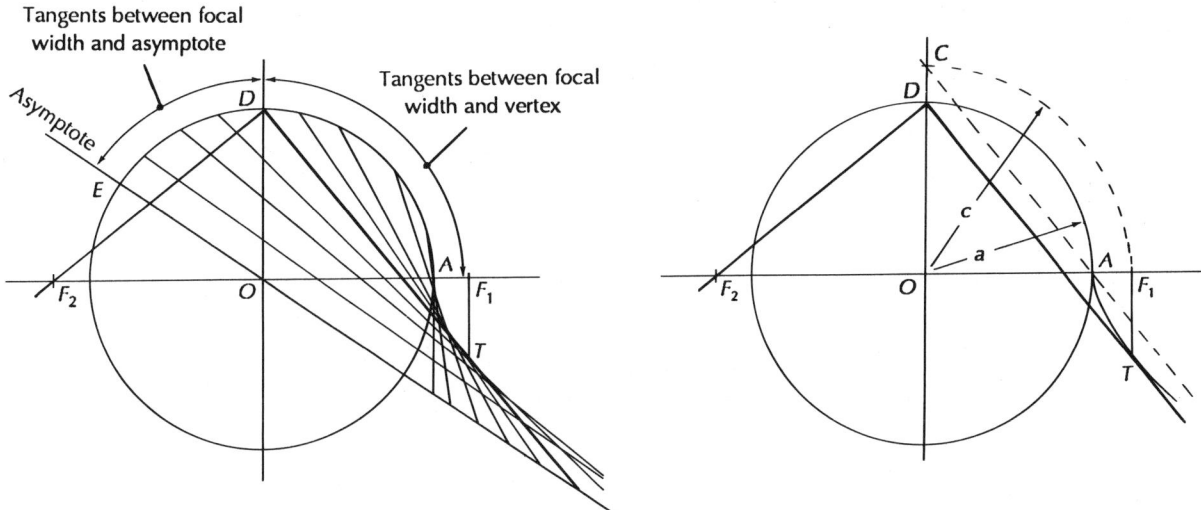

FIGURES 4-3 AND 4-4

To prove that the tangent we have generated at the focal width is equal to its eccentricity, we can use the same method we used for the ellipse—that is, to construct a line parallel to the tangent but passing through the vertex at point *A*. Line *CA* has been constructed parallel to line *DT*. Since *OC* is equal to the dimension c and *OA* is equal to a, we have $e = \dfrac{c}{a}$, which is both the eccentricity of the curve and the slope of the tangent at its focal width.

As we shall see in some detail in Chapters 5 and 6, an ellipse must have tangents less than unity (45°) at its focal width. Otherwise a ray reflected from that point would not be directed in a manner that would cause it to cross the major axis at its second focus. Rays reflected from the surface of an ellipse are convergent (Figure 4-5).

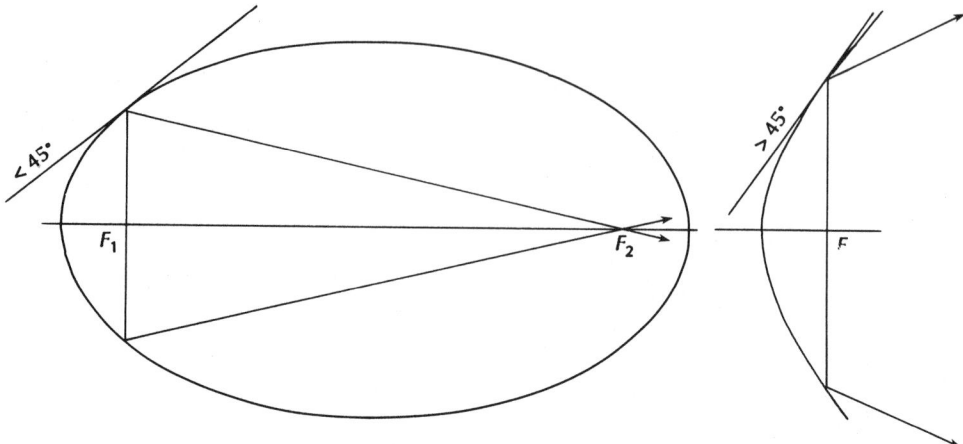

FIGURES 4-5 AND 4-6

Likewise, a hyperbola at its focal width must have a tangent that will cause rays to be reflected in a divergent manner, appearing to have originated at the other focus. Hyperbolas must have a tangent greater than unity to do this (Figure 4-6).

CONSTRUCTING A PARABOLA

With ellipses having tangents less than unity and hyperbolas having tangents greater than unity, we can justifiably ask what happens when the tangent is unity. We know that the tangent of a parabola at its focal width is unity (tangent of 45°), but when we use the Directing Circle method—as in the case of the hyperbola and ellipse—we find that as the foci approach the circle, the curve becomes so flat that at the circle there is no vertical dimension.

Do not despair. Our problem is that the Directing Circle must be infinitely large to generate a parabola. As we explained before, a straight line represents a portion of an infinitely large circle. Admittedly, when we use infinity as a dimension, we are in a gray area, since the radius of a circle whose diameter is infinite is also infinite. Since this book presents the practical aspects of conic curves, we shall leave such theoretical niceties to the theoreticians.

Finding the exact point of tangency on a parabola is quite simple. After establishing the axis, focal point, and vertex line, place the right angle of your triangle at any point along the line that is tangent to the vertex. With one arm of the triangle passing through the focus, the other arm will be tangent to the parabola. The exact point of tangency can be found by constructing a perpendicular at a point that is twice the distance from the right angle to the axis, as shown in Figure 4-7.

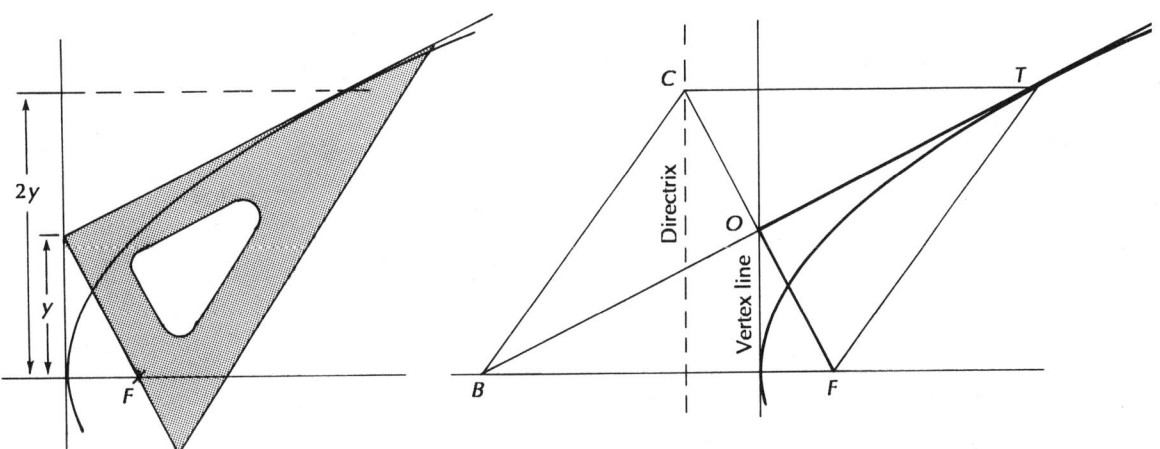

FIGURES 4-7 AND 4-8

The proof of this fact is equally simple. By extending some of the lines in the previous drawing, we find that we have constructed a rhombus with diagonals dividing it into four congruent triangles. Note that in all cases $FO = OC$. Any line passing through the focus will cross the vertex line at a point exactly half the distance to the directrix. Diagonals of a rhombus are perpendicular; in fact, the right triangle FOT is all that we use to generate the parabola; the remaining parts are needed only for proof (Figure 4-8).

POINTS OF TANGENCY

The directrix of the parabola is a straight line perpendicular to the axis and is located behind the vertex at a point equal to the distance from the vertex to the focus. This should be remembered because a "directrix" of sorts is used to locate the exact point of tangency in ellipses and hyperbolas.

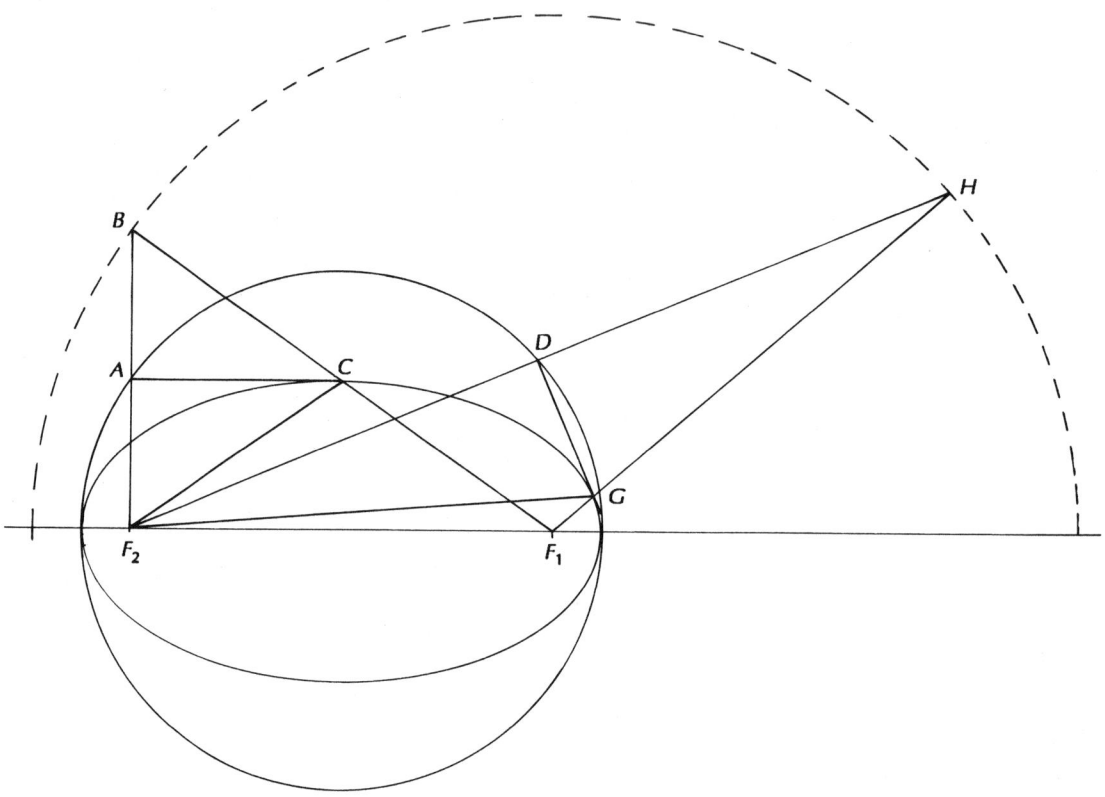

FIGURE 4-9

To locate the point of tangency on an ellipse, draw a line from one focus through the Directing Circle to a point exactly twice the distance from the focus to the Directing Circle. Draw a line from that point to the other focus (Figure 4-9). Where this line BF_1 crosses the ellipse at C is the point of tangency. Since $AB = AF_2$, it is obvious that $\triangle$s ACF_2 and ACB are congruent. It is also clear that $\triangle\ ACF_2$ is really the same thing as the drafting triangle used to generate the ellipse, since its right angle always occurs on the Directing Circle.

If we repeat the operation a number of times (using the triangle F_2HG for example), we find that the locus of the point that is twice the distance from the focus F_2 to the Directing Circle is also a circle whose center is at the other focus (F_1) and whose radius is equal to the major axis of the ellipse. This circle serves the same purpose as the directrix of the parabola in Figure 4-8.

FIGURE 4-10

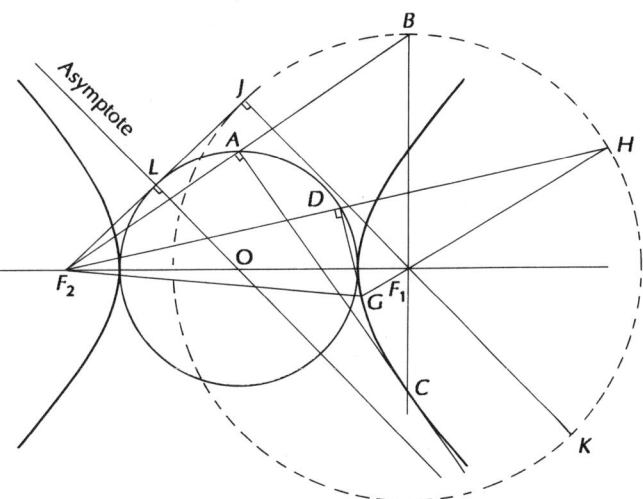

The exact point of tangency to a hyperbola may be found in a similar manner. Draw the line F_2A (Figure 4-10) passing through the Directing Circle and extending to point B, which is twice the distance from F_2 to A. Draw the line BC, passing through F_1. The resulting isosceles triangle, ΔF_1BC is divided into two congruent right triangles by the line AC. One of these triangles (ΔF_2AC) is, in fact, our old friend the drafting triangle used to generate the hyperbola in the Directing Circle method of construction.

If we repeat this operation a number of times (using, for example, ΔF_2HG), we find that the locus of the point that is twice the distance from F_2 to the Directing Circle is also a circle whose center is at the other focus (F_1) and whose radius is equal to the distance between vertices of the pair of hyperbolas. (This distance is usually referred to as $2a$ and is the length of the major axis.) The new circle serves the same purpose as the directrix of the parabola in Figure 4-8. Note that the line $F_2 J$ is tangent to both circles. The perpendicular at L passes through the center of the Directing Circle, becoming an asymptote, and the perpendicular at J passes through F_1, which is the center of the other circle. The line JK is parallel to the asymptote.

COMPARING CONIC CURVES

When we find a geometric attribute in ellipses, we can usually find a related attribute in hyperbolas, but reversed. This is to be expected when you consider that ellipses are generated inside the Directing Circle and hyperbolas are generated on the outside. In Chapter 5 we shall explore the reflective properties of ellipses and hyperbolas, both concave and convex, which show a relationship as predictable as that found in the construction of these curves.

At this point we might be tempted to conclude that ellipses are the mirror images of hyperbolas and that for every ellipse there is a corresponding but opposite hyperbola. This is not true. There are basic differences between them. Ellipses are closed, finite figures, but hyperbolas are open and infinite. There are no minor axes in hyperbolas and no asymptotes in ellipses, and not all construction techniques for one figure have counterparts for construction of the other—for example, the trammel method works only for ellipses.

As mentioned earlier, either focus of the ellipse and the hyperbola may be used for construction, but the focus farther away from the vertex of the curve being drawn expands the amount of the Directing Circle used and is preferred when constructing small figures. In practice, when constructing, for example, a 10-foot hyperbolic reflector for a large radio telescope, it is much more practical to use the focus that is closer to the vertex. Figure 4-11 shows an ellipse, a parabola, and a hyperbola, each with the same focal distance from the vertex but constructed using the nearer focus.

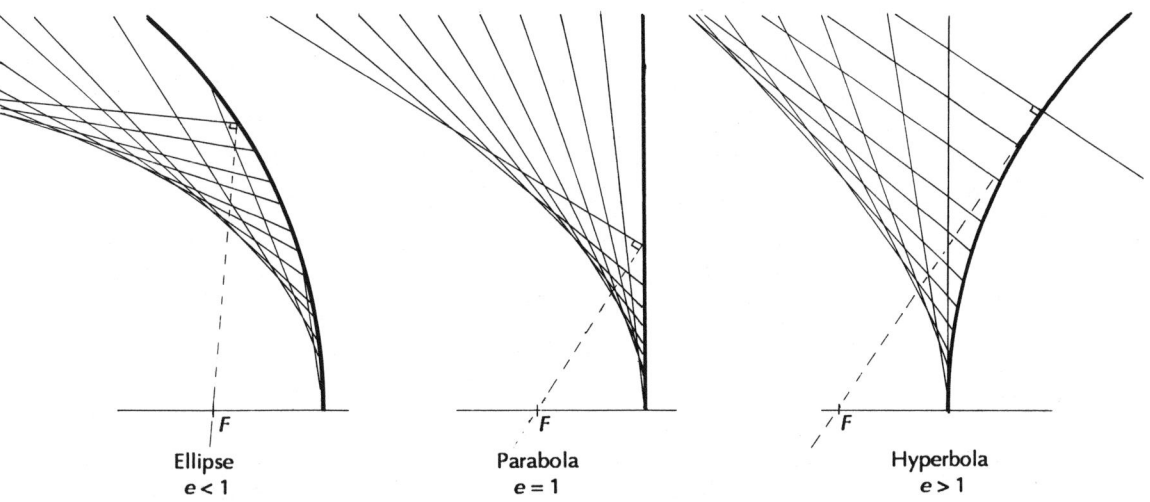

FIGURE 4-11

This graphically shows the relationship of the curves. It also makes it obvious that in constructing hyperbolas, an asymptote is reached when the arm of the right angle that passes through the focus is tangent to the Directing Circle. Beyond that point the lines generated are no longer tangent to that hyperbola but are tangent (when extended) to the hyperbola on the other side of the Directing Circle.

CHAPTER FIVE

REFLECTIVE PROPERTIES OF SOLID CONIC CURVES

Each conic curve has its own characteristic ability to reflect radiant energy and is the only figure that can do its particular job. In this discussion we divide these curves into two groups: those that are open and those that are closed. We shall assume in our discussion that each conic section curve has been rotated about its axis to form a solid. Thus a circle becomes a sphere, an ellipse becomes an ellipsoid, a parabola becomes a paraboloid, and a hyperbola becomes a hyperboloid.

ELLIPSOIDS

An ellipse is formed when a plane cuts completely across a right circular cone. The resultant figure has a center, two foci, a major axis, and a minor axis. An ellipse may be rotated about either axis to form an ellipsoid. When an ellipse is rotated about the major axis, the resultant football-shaped figure has two foci, but when the ellipse is rotated about the minor axis, the foci form a single circle within the ellipsoid. Of these two figures (known as *prolate* and *oblate spheroids,* respectively) the former is by far the more useful and is the only one to be considered in this discussion of reflective properties. However, it should be noted that the earth is an oblate spheroid, since its polar diameter is less than its equatorial diameter.

Ellipsoids have the ability to reflect energy radiating from one of their focal points to the other focal point. Not only is the reflective angle at any point on the inside surface correct for reflecting energy from one focus to the other, but also the total distance covered is constant. (Remember that an ellipse is defined as the locus of a point whose distance to two fixed points is a constant.) This means that wave energy, such as sound, ultrasonic, light, or microwave energy, will arrive at the second focus after having traveled exactly the same distance regardless of which path it took; it will also arrive in phase with energy that took a different path (Figure 5-1).

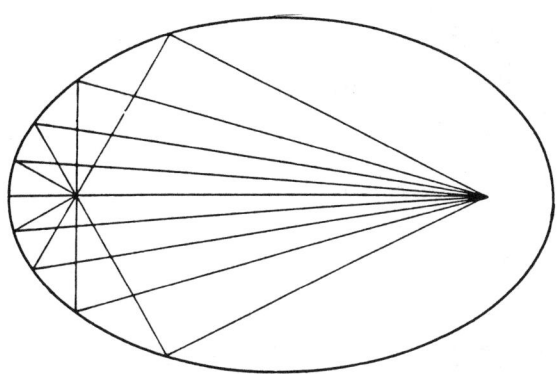
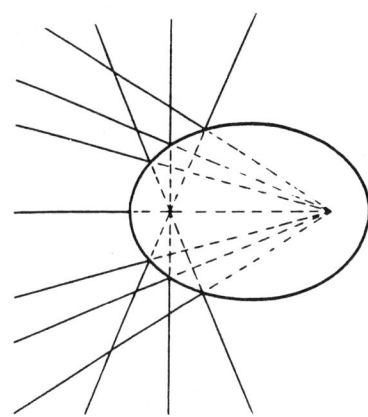

FIGURES 5-1 AND 5-2

A reflective property frequently overlooked in an ellipse is that radiant energy converging toward one focus, upon reflection from the external (convex) surface, emerges in a divergent ray system in which the other focus is the apparent point of origin. Figure 5-2 shows that reflected energy appears to have originated at the other focus regardless of which focus is used as the target.

An ellipsoidal reflector can collect a widely divergent ray system and concentrate it into a cone of much smaller angular aperture. A condenser lens may then be placed after the second focus to utilize the concentrated image of the light source rather than the light source itself. Reflectors like this are frequently used in motion picture projectors (Figure 5-3).

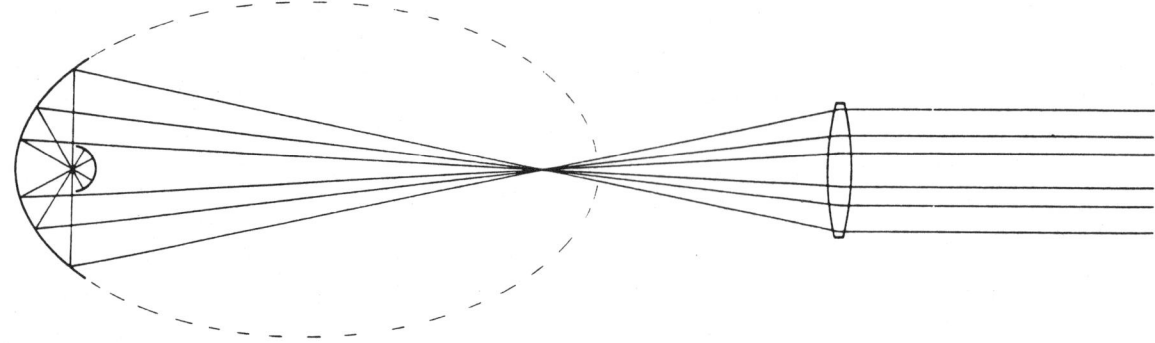

FIGURE 5-3

All the observations made about an ellipsoid may be made to apply to a sphere. A sphere is a special ellipsoid whose two foci are coincident and whose major and minor axes are equal. Radiant energy originating at the focus (center) will be returned to the same point regardless of which path it took. Since all diameters of a circle are equal, a sphere may be generated by rotating it about any line passing through its center. Spherical reflectors are of little practical use, since there are few applications that demand the simple return of radiant energy to its point of origin. However, in a situation where it is desirable to exclude radiation from a sector, it is possible to utilize a portion of a sphere. This portion will reflect energy back through the source to join the energy that is radiated directly on the side opposite the reflector, as shown in Figure 5-3.

Ellipsoids have traditionally been generated by rotating ellipses about one of their two axes. We are by no means limited to these; we may rotate ellipses about any line we chose. Ellipsoids rotated off-axis retain their basic reflective properties and allow us to use a focus of finite length instead of an infinitesimal point. Figure 5-4 shows an ellipse rotated about a line that is perpendicular to the major axis and passes through one focus. The resulting figure resembles a malformed donut having a point focus at F_2 and a distributed focus at F_1, which has become a circle with a radius equal to the distance between foci. Energy leaving this circle (perpendicular to the circle) will be reflected to F_2. The importance of this is that we can get an enormous concentration of energy from a distributed source to a theoretically infinitesimal point.

FIGURE 5-4

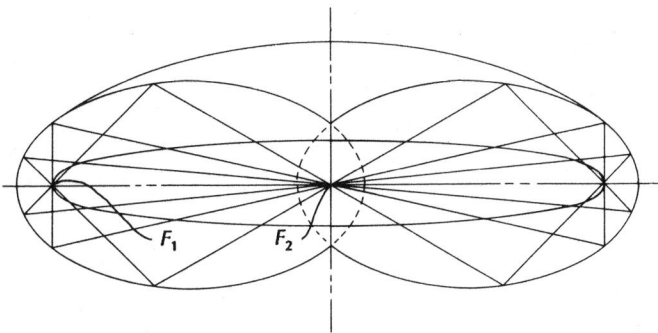

This use works well with ultrasonic energy and explosive wires, which may be placed along the distributed focus. Such energy leaves perpendicular to the source and strikes the surface of the ellipsoid in the proper plane to be reflected to F_2 (Figure 5-5). Unfortunately, light energy is not so well disciplined, and each point along a filament or gas-discharge tube radiates its energy in all directions, as shown in Figure 5-6. This introduces a large axial error for much of the energy if one tries to use this configuration to concentrate light. However, an experimental model using a 5-inch diameter gas-discharge tube has been used to concentrate an impressive amount of energy at the point focus. Even if it were possible to concentrate all the light energy from such a source, there is an immutable law of nature that does not permit the temperature of an image of incoherent light to exceed the temperature of the source. Laser light, being coherent, is not bound by this law and may reach temperatures high enough to bring about atomic fusion.

FIGURES 5-5 AND 5-6

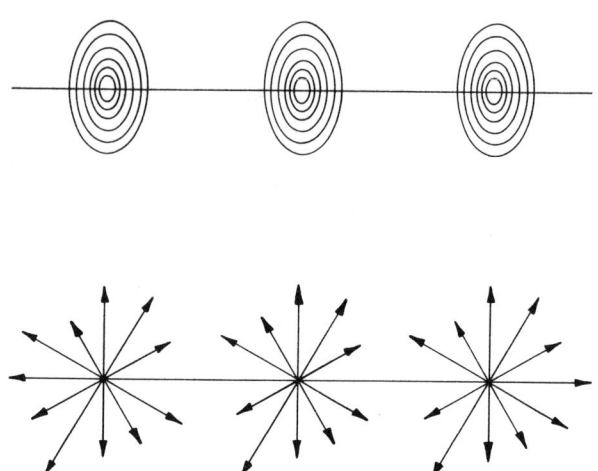

Axes of rotation are not limited to the example shown. Figure 5-7 illustrates an ellipsoid that has been formed by rotating an ellipse about a line crossing the major axis at one focus at an acute angle. The classical rules of reflection within an ellipse will be observed and energy will converge on the point focus as it did before, but at the apex of a cone.

FIGURE 5-7

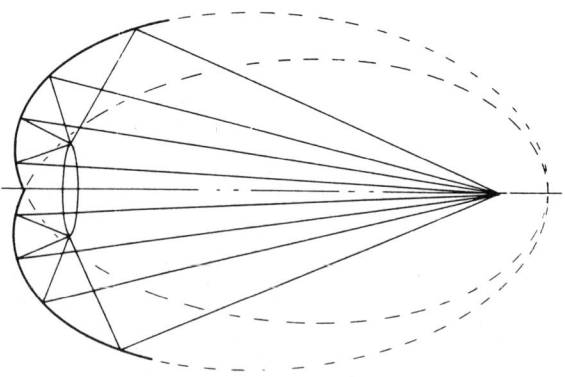

Figure 5-8 shows a group of three elliptical cylinders that share a common focus. Elliptical reflectors like this have been used to excite solid lasers. Gas-discharge flashtubes are placed along the line focus of the three ellipses. When the flashtubes are fired, all the energy converges on the common focus, exciting the laser to fire. Large concentrations of energy are possible in such an arrangement.

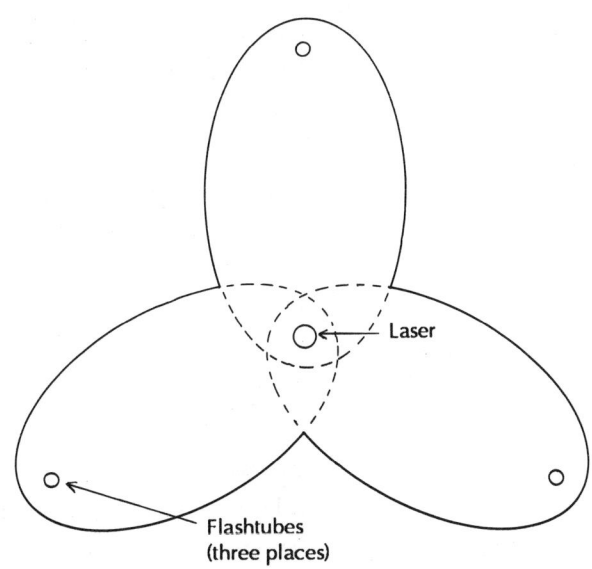

FIGURE 5-8

Elliptical reflectors for exciting a laser.

HYPERBOLOIDS

As we might expect, hyperboloids have unique reflective properties that may best be described as similar to those of ellipsoids but turned inside out. Radiant energy originating at the focus of a hyperboloid will, after reflection, diverge with the focus of the other hyperboloid (imagined) as the apparent source (Figure 5-9).

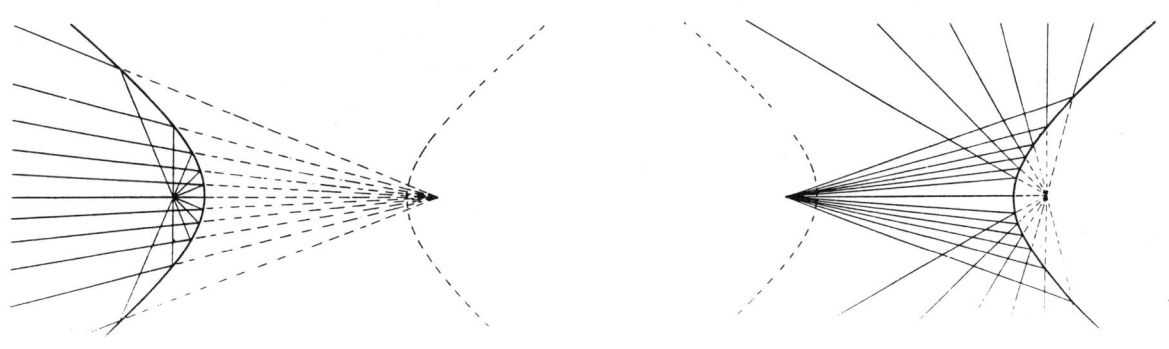

FIGURES 5-9 AND 5-10

Conversely, when radiant energy is converging toward the focus of a convex hyperboloid, it will, after reflection, converge on the other focus (Figure 5-10). This is the most frequently used reflective property of hyperboloids and is the basis for Cassegrain telescopes and microwave antennas. This property will be discussed in Chapter 6.

It was pointed out in Chapter 2 (Figure 2-6) that a hyperbola whose constant is zero is a line midway between the foci and perpendicular to the line drawn through the foci. A hyperboloid generated from such a "hyperbola" is a flat plate. Figure 5-11 shows that the reflective properties still hold true for this special hyperboloid. In fact, these properties have been used in some Cassegrain microwave antennas, although the blockage of the incoming energy is excessive due to the relatively large size of the flat secondary reflector compared to the more conventional hyperboloids.

FIGURE 5-11

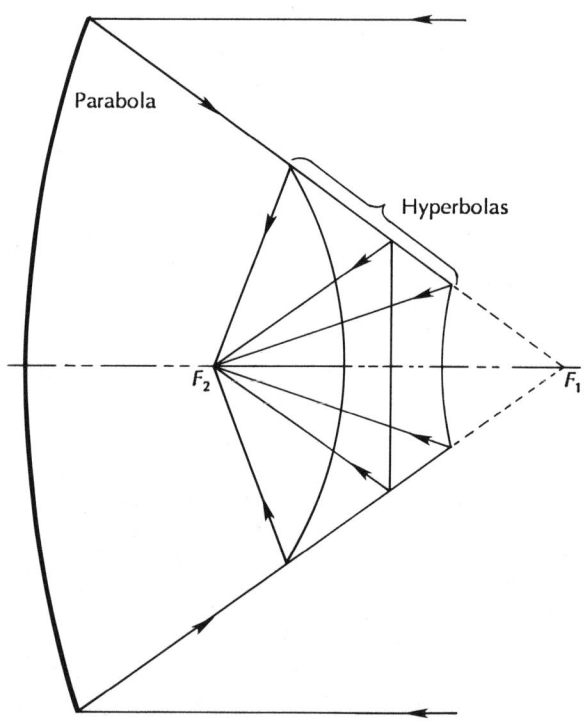

42 | PRACTICAL CONIC SECTIONS

PARABOLOIDS

Since the parabola is a special case of both the ellipse and the hyperbola, we would expect to find the reflective properties of paraboloids to be essentially the same as those of ellipsoids and hyperboloids. They are, in fact, exactly the same (concave and convex) if we consider that the second focus in each case is infinitely distant from the reflector. Rays that converge on a point that is infinitely distant are parallel.

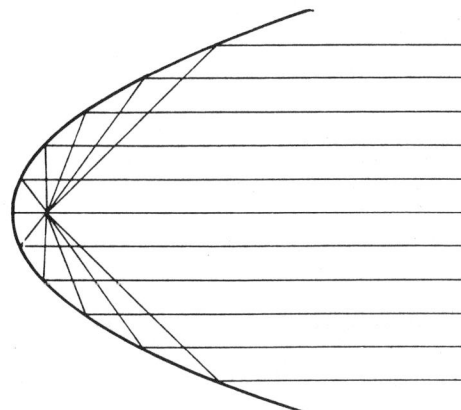
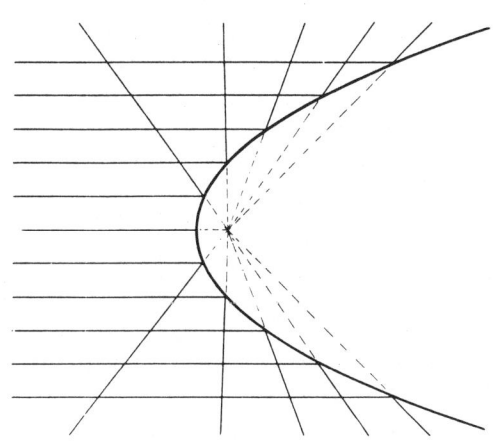

FIGURES 5-12 AND 5-13

Without doubt, paraboloids are the most widely used of all reflective devices based on conic curves. They find application in the projection and gathering of light, radio, and sound energy and range in size from hand-held flashlights to radio telescopes.

Although it is customary to use symmetrical parabolic reflectors whose vertex is at the center of the dish, we are not limited to this configuration. The laws of reflection from the surface of a paraboloid apply to all parts of the curve. Any portion of the curve may be used as long as the receiver or source of energy (light, radio, sound, and so on) is at the focus (Figure 5-14). The advantage of an offset feed is that the source will not be in the way of the energy path. The disadvantage is that the efficiency of utilization of the reflective surface area decreases rapidly when we get beyond the focal width of the paraboloid. At the focal width, the efficiency is reduced to 70.7%. This may not be a significant problem with small reflectors for radar and optics, but it becomes quite serious when we consider large parabolic reflectors used for concentrating solar energy. Such reflectors would require approximately 50% more reflective material at the focal width than at the vertex to intercept a given amount of incident sunlight. Because of this, it is best to design reflectors to be rather shallow, having a diameter less than four times the focal length.

FIGURE 5-14

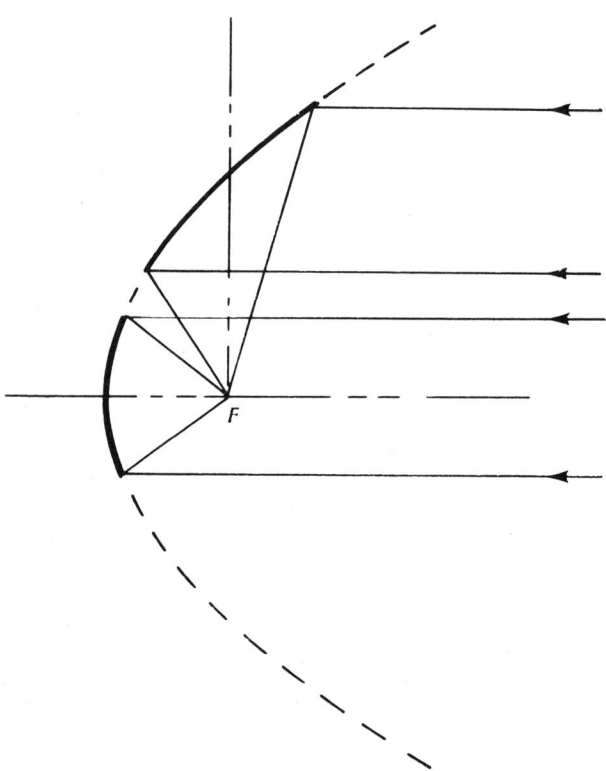

The effective focal length of a deep paraboloid is twice as long at its focal width as it is at the vertex. Since the size of the projected image is directly proportional to the effective focal length, the image size projected from the area of the vertex will be only half the size of the image projected from the area of the focal width. Although this probably will not cause any problems in the projection or gathering of energy, it is a phenomenon of which we should be aware.

Parabolas may be rotated off-axis in the same way as the ellipses. The result works better for incoming energy concentration than as a source of projected energy. The advantage of this configuration is that solar energy can be concentrated on to a circular collector of finite length instead of a small but intensely hot spot (Figure 5-15).

FIGURE 5-15

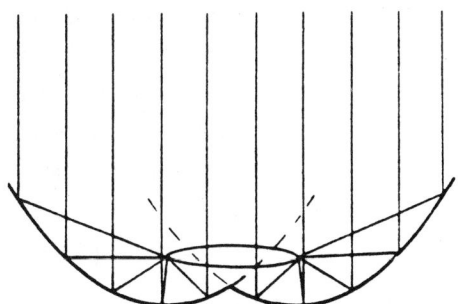

CHAPTER SIX

COMPOUND REFLECTORS

Compound reflectors are reflectors that use two or more reflective surfaces. Usually these reflectors involve a primary parabolic reflector and a secondary hyperbolic or elliptical reflector to make a Cassegrain or Gregorian system.

CASSEGRAIN SYSTEMS

The Cassegrain principle was invented by N. Cassegrain, a French physician and inventor, in the seventeenth century. Its first use was in astronomical telescopes, but it has since been extended to tracking and communications antennas. It consists of a primary parabolic reflector and a secondary hyperbolic reflector sharing its focus. Two reflectors having a common focus are said to be *confocal* (Figure 6-1).

FIGURE 6-1

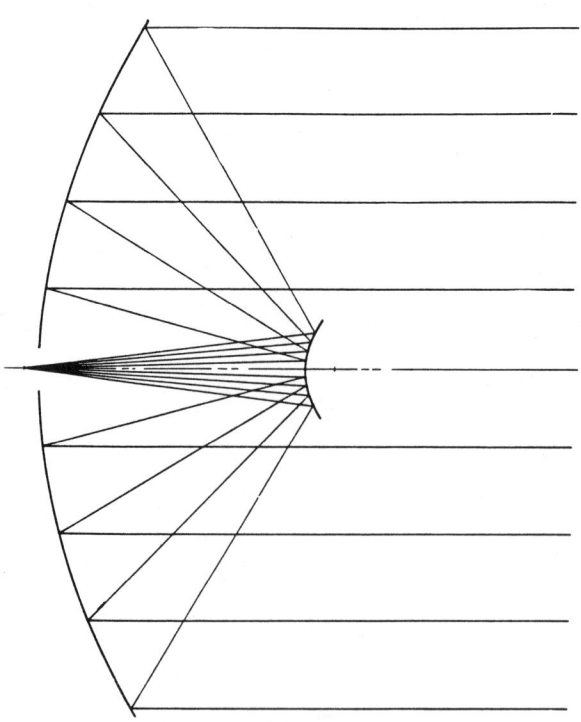

As light or radio energy reflected by the primary parabolic reflector approaches the focus, it is intercepted by the hyperbolic secondary reflector and is sent to the second focus of the hyperboloid.

This system has several advantages. The observer is now looking through an astronomical telescope tube in line with the object being observed instead of observing through a hole in the side of the tube, as is the case with Newtonian telescopes. The effective focal length of a Cassegrain telescope is usually longer than the tube housing the reflectors. The telescope in Figure 6-2 would have to have a tube two or three times the length shown if it were a Newtonian telescope or refractor of the same power.

FIGURE 6-2

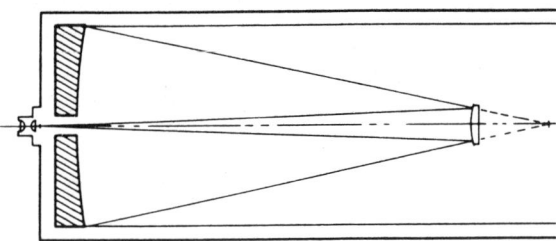

The disadvantage of the Cassegrain telescope is in the construction. Alignment and accuracy of the reflectors must be more precise than required by the simple primary-focus (Newtonian) telescopes. Alignment is so critical that in Albert Ingall's book *Amateur Telescope Making* there is a chapter titled: "How to Make a Cassegrain Telescope (and why not to)." The disadvantages are not insurmountable, and many fine Cassegrain telescopes have been constructed in sizes ranging from the small 4-pound Celestron used as a telephoto camera lens to the 200-inch-diameter telescope at Mount Palomar.

Cassegrain microwave antennas work on exactly the same principle as the telescopes, but the antennas are usually much larger and deeper than astronomical telescopes. Again, there are advantages and disadvantages to this system. It is usually helpful to have the feed horn back near the vertex of the primary reflector, requiring less waveguide or coaxial cable than by placing it at the primary focus, but the feed horn must be large to illuminate the relatively small secondary (hyperbolic) reflector (Figure 6-1).

GREGORIAN SYSTEMS

Gregorian systems (Figure 6-3) are very similar to Cassegrains, and the results are identical. The primary (parabolic) reflector is confocal with an ellipsoidal reflector. The primary reflector passes its energy through its focus and onto the surface of a concave ellipsoid, which reflects the energy to the second focus of the ellipsoid.

FIGURE 6-3

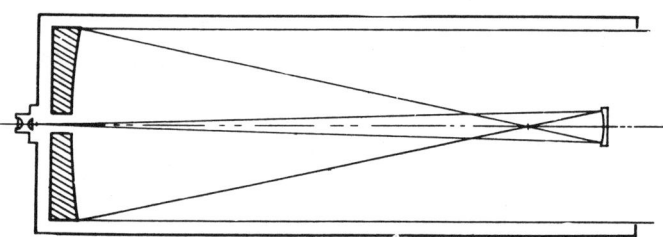

Although Gregorian telescopes function as well as the Cassegrains, they are not widely used because of the deeper curve of the ellipsoidal reflector and the added tube length of the system.

USING THE CASSEGRAIN PRINCIPLE

The Cassegrain principle may use the concave as well as convex reflective properties of hyperbolas. Figure 6-4 shows three secondary reflectors, which are confocal with the primary (parabolic) reflector. For practical purposes, the convex secondary reflector is the only one that can be taken seriously, since the diameter of the others increases to the point that blockage of the primary reflector becomes significant, although flat secondary reflectors are sometimes used.

FIGURE 6-4

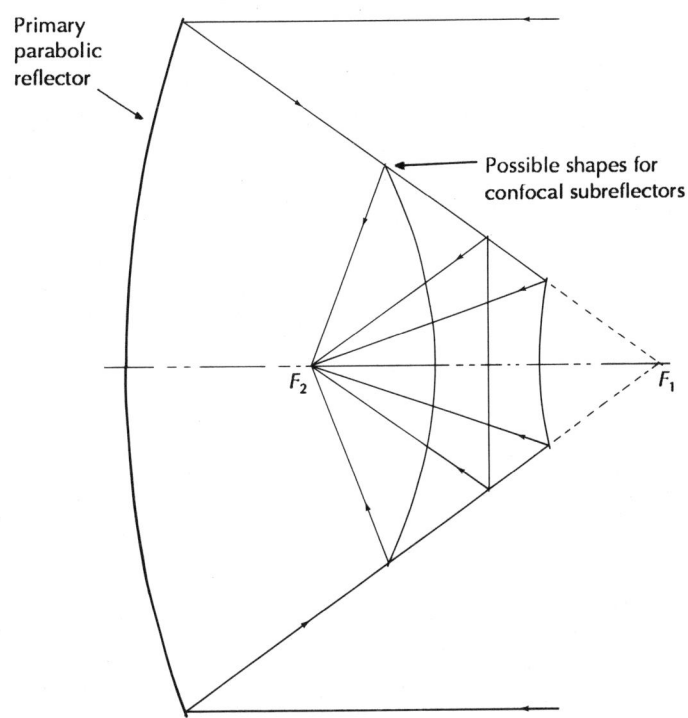

The Cass-horn antenna (Figure 6-5) is a modified form of the Cassegrain antenna. Although it involves a hyperboloid, it uses its properties to reflect as though the energy had originated from a point outside the physical confines of the antenna. This concave hyperboloid is confocal with the primary parabolic reflector and is one of the few known uses of the internal reflective properties of hyperboloids. This system has not proved to be practical and has fallen from favor during the past twenty years.

FIGURE 6-5

Cass-horn antenna.

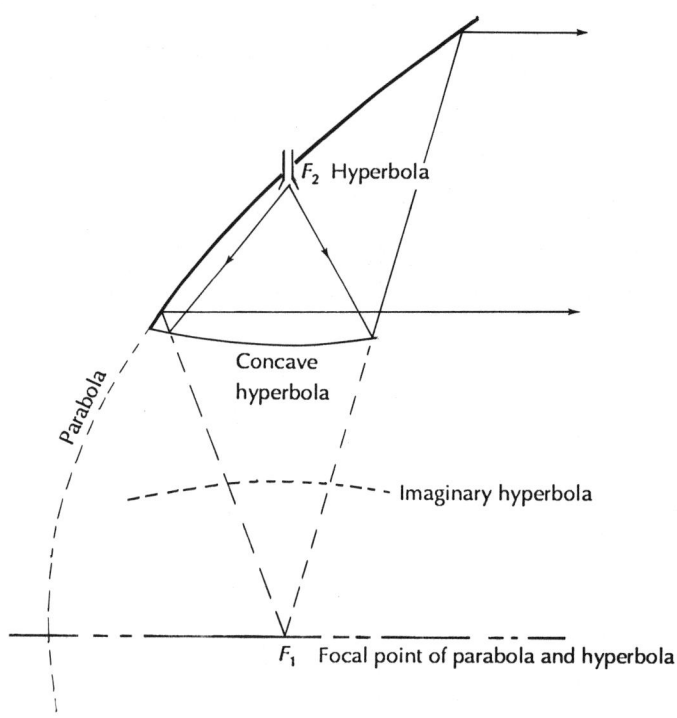

COMPOUND REFLECTOR DESIGN

The Directing Circle method of generating conic curves that was described in Chapters 1–4 can be extremely useful in determining the size and shape of a secondary reflector in a Cassegrain or Gregorian system.

As an example, let us derive the curves involved in a Cassegrain radio telescope consisting of a 20-foot parabolic reflector having a focal length of 6 feet. The hyperbolic secondary reflector is to have a diameter of 4 feet and a distance between foci of 5 feet. Using a convenient scale such as 1 in. = 1 ft, draw the axis of the system and construct a perpendicular at point A. This line is the tangent to the vertex of the primary reflector.

FIGURE 6-6

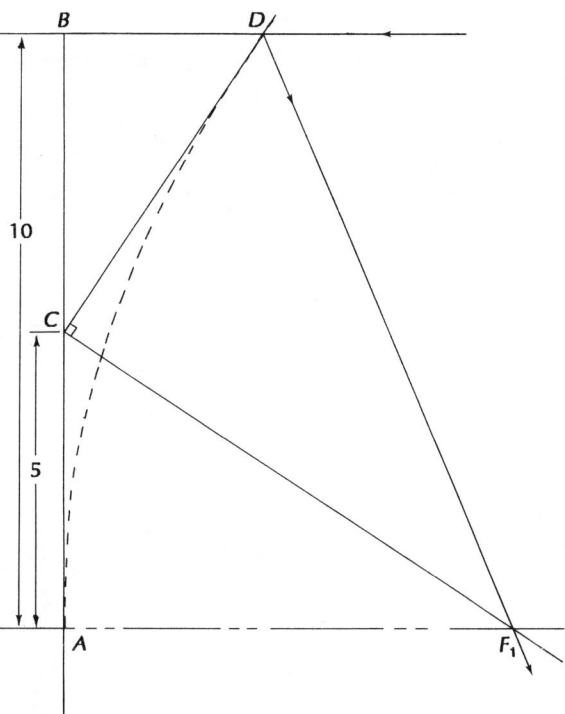

Measure 6 units from point A and establish the focus F_1 (Figure 6-6). Draw a line parallel to the axis through point B 10 units (representing the radius of the 20-foot reflector) above the axis. Bisect the line AB to establish the midpoint C. Using a drafting triangle, place the right angle at C with one edge passing through F_1. The other edge of the right angle will be the tangent of the parabolic curve at the edge of the reflector at point D. We now know all we really need to know about the primary (parabolic) reflector—its diameter, depth, and focal point.

In Figure 6-7 we have drawn two parallel lines 2 units on either side of the axis. These represent the edge of the 4-foot diameter hyperbolic subreflector. The tangent of the hyperboloid at point E must be at the exact angle needed to intercept a ray aimed at F_1 and reflect it toward F_2, which we have established 5 units from F_1. Since the angle of reflection is equal to the angle of incidence, the tangent at this point will be the bisector of ∠ DEG, which we construct geometrically and label JK.

FIGURE 6-7

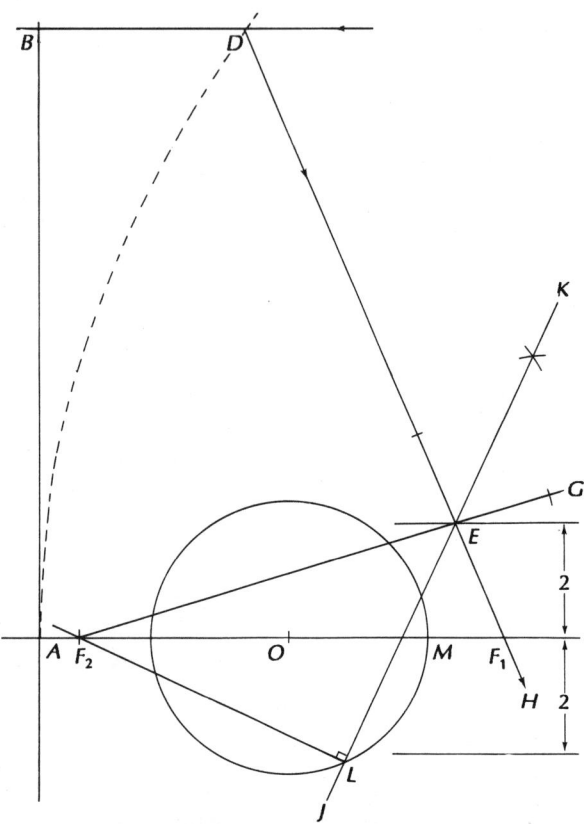

Construct a perpendicular from F_2 to line JK. (This may be done in the classical manner with compass and straightedge or, more practically, by positioning a drafting triangle with one edge directly over line JK with the other edge passing through F_2.) Point L is now established as a point on the circumference of the Directing Circle as explained in Chapter 2. The center (O) of the Directing Circle may be found by bisecting the line F_1F_2. We may now draw the Directing Circle, since we have the center and a point known to be on the circumference. The vertex of the hyperbola occurs at point M, where the circle crosses the axis. We now know the depth and location of the secondary reflector, which may be all that we need to know at this time if we are simply sketching the system. At this point it is easy to juggle any of the variables (the distance between foci and the diameter of the subreflector) to obtain a visual concept of the shape and spacing of the two reflectors. Once we have established exactly what we want, we may complete the hyperbola by drawing a representative number of tangents, as shown in Figure 6-8.

FIGURE 6-8

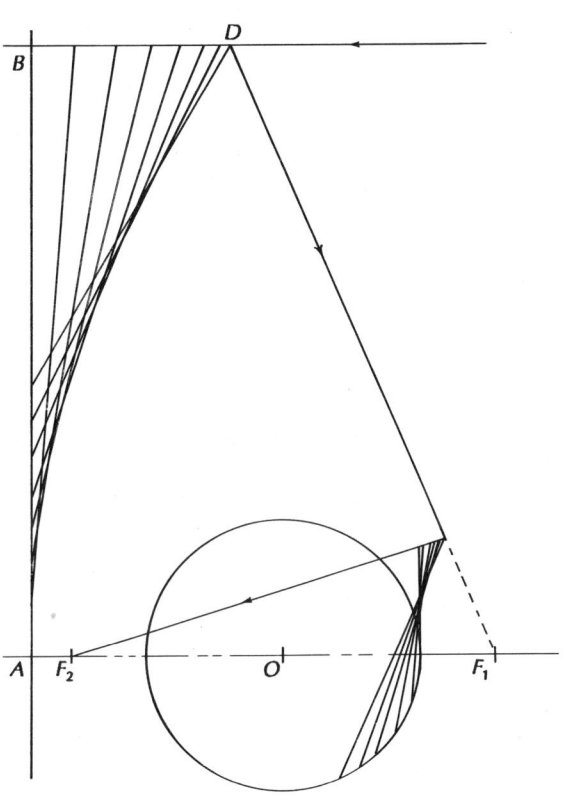

Although Gregorian compound reflectors are not as popular as Cassegrains, there are some applications in which they are desirable. They can be used to achieve exactly the same function but with an ellipsoidal reflector beyond the focus of the parabolic primary reflector. To design a Gregorian compound system having the same design parameters as the previous example, start with the same 20-foot primary reflector and the same placement of the feed horn, but this time make the secondary reflector surface an ellipsoid located beyond the primary focus, as shown in Figure 6-9. Using the Directing Circle method outlined in Chapter 1, it is as easy to lay out a Gregorian system as it is to do a Cassegrain.

FIGURE 6-9

Gregorian antenna.

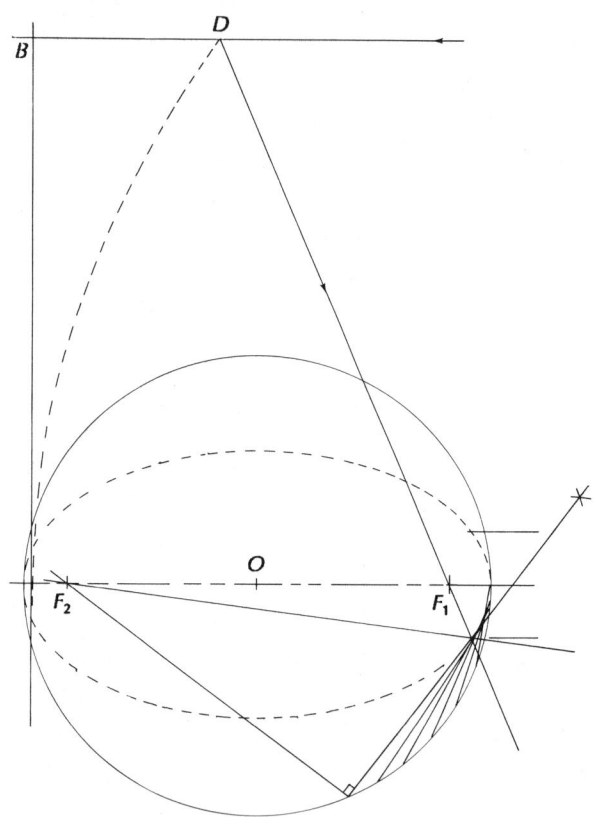

CHAPTER SEVEN

ECCENTRICITY

In our discussions of the conic section thus far, we have briefly mentioned the *eccentricity* of the conic curves. However, no discussion of the construction of conic curves would be complete without an explanation of eccentricity and the special cases of conic curves. Eccentricity is defined as the distance between foci divided by the distance between vertices, which may be stated mathematically as $e = \dfrac{c}{a}$.

Figure 7-1 shows the values of e for conic curves. When the distance between vertices ($2a$) is greater than the distance between foci ($2c$), the resulting figure is closed. When the distance between foci is greater that the distance between vertices, a pair of hyperbolas results.

FIGURE 7-1

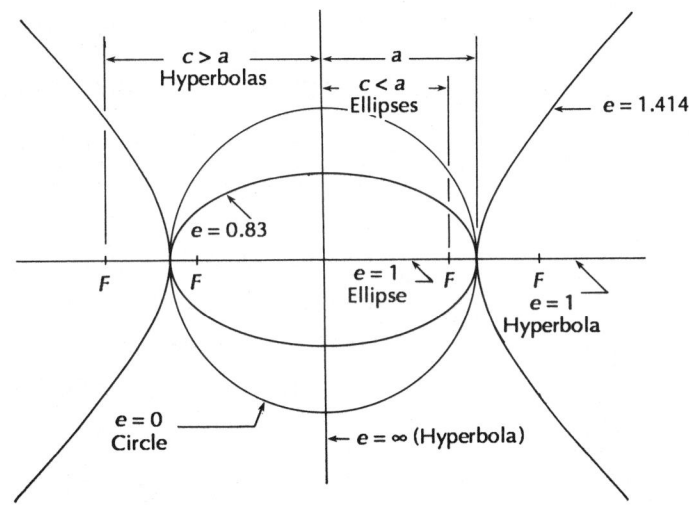

It is sometimes easier to think of eccentricity as the displacement of the foci from the center as a percentage of *a*. For example, the sun is displaced from the center of the earth's elliptical orbit by 1.67%, which works out to be slightly over 1.5 million miles. A parabola has an eccentricity of 1, so its focus is placed 100% of the distance from the center, which is infinitely distant. Hyperbolas have eccentricities greater than unity, so the corresponding percentages will be greater than 100%. In rectangular hyperbola, for example, the foci are displaced from center by 141.4%.

SPECIAL CASES

There are four examples of eccentricity that are of special interest:

$e = 0$ (circle)

$e = 1$ (parabola)

$e = \sqrt{2}$ (rectangular hyperbola)

$e = \infty$ (vertical line for hyperbola, horizontal for ellipse)

If we set the focal distance for e to be 0, the foci are placed at one point in the center of the ellipse. Therefore, $e = \dfrac{c}{a} = \dfrac{0}{a} = 1$. The major and minor axes are equal, and the resulting "ellipse" is a circle. The formula for the area of an ellipse is $A = \pi ab$, where a is half the length of the major axis and b is half that of the minor axis. Since a circle is a special case of an ellipse in which the major and minor axes are equal ($a = b$), the area formula may be written as $A = \pi r^2$. Unfortunately, mathematicians have not devised such a neat relationship for the circumference of an ellipse.

Referring again to Figure 7-1, when the focal distance c is equal to the distance to the vertex, $e = \dfrac{c}{a} = 1$. The results will be either a straight line the length of the major axis ($2a$) of an ellipse or two straight lines moving away from the ends of the line $2a$; the latter case may be considered to be a pair of hyperbolas outside the Directing Circle. Either way we look at these figures, they have only length and no second dimension.

Again, we find that the conic curve with an eccentricity of 1 is the parabola; as shown earlier, we cannot give a finite dimension to either the major axis or the focal distance as measured from the center of an infinitely large circle. If these are both infinite, we have $e = \dfrac{c}{a} = \dfrac{\infty}{\infty} = 1$. Mathematically speaking, we are in a gray area, but from a practical standpoint, we are given license to place the focus anywhere we wish on the axis. The line tangent to the vertex will really be a straight line perpendicular to the axis, but it may be considered to be a portion of a circle of infinite diameter. This is the reason that the focal length of a parabola is always measured from the vertex, whereas the foci of ellipses and hyperbolas are usually measured from the point midway between foci.

Hyperbolas having an eccentricity of $\sqrt{2}$ are interesting and somewhat useful. These hyperbolas are called *rectangular,* or *equilateral,* and their asymptotes are perpendicular to each other. Figure 7-2 shows a set of four such hyperbolas whose foci have been joined to form a square. Since this square exactly encompasses a circle whose radius is the distance *a,* the diagonals of this square must be $\sqrt{2}$ times the length of a side of the square.

FIGURE 7-2

Rectangular hyperbolas.

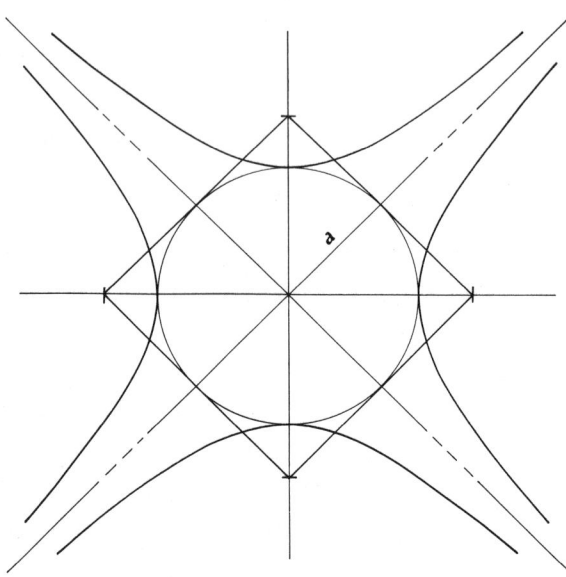

Rectangular hyperbolas follow the formula $PV = k$, where $P =$ pressure, $V =$ volume, and k equals some constant. They may be used to represent graphically the relationship of pressure and volume of a gas when the temperature is constant (Figure 7-3).

FIGURE 7-3

Rectangular hyperbola representing the relationship of pressure and volume: $PV =$ constant.

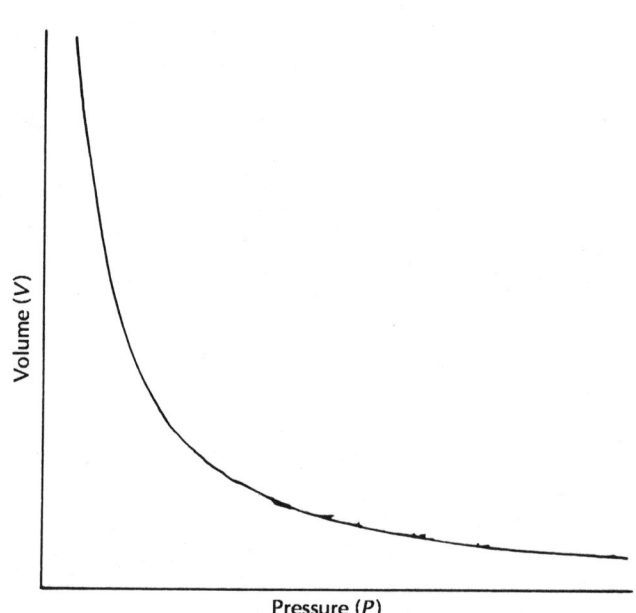

As the eccentricity increases, hyperbolas become more obtuse, until they become straight lines at infinite eccentricity. The easiest way to reach infinity for practical purposes is to divide by zero. To do this we set the distance a at 0 while leaving c any finite value. This gives us $e = \frac{c}{0}$. As shown in Figure 2-6, we have no distance between vertices, and the resulting "curve" is a straight line perpendicular to the axis. This "hyperbola" follows all the construction and reflective properties of the other hyperbolas.

Some of those who have taken a class in first-year analytical geometry probably emerged with little more than a numerical concept of apparently unrelated conic curves and a possible concern about how one can get a large enough piece of graph paper to plot the coordinates of planetary orbits. However, we can think of this marvelous set of curves as a continuous family. Visualize the foci starting at the center of the Directing Circle, moving symmetrically out as the ellipse pulls away from the top and bottom of the circle. As the foci approach the circle, the ellipse flattens until, at the circle, the ellipse is flat. As the foci move outside the circle, the resulting hyperbolas are quite acute, but as the foci move farther out, the hyperbolas become more obtuse, until—at distant points—they become nearly flat.

CHAPTER EIGHT

FOR PRACTICAL PURPOSES

So far in our discussion we have treated conic curves as perfect entities with perfect axes and focal points. This purity of geometry is nice on paper, but anyone who has ever adjusted a telescope or a radar antenna knows that these lines and points are not as inflexible as theory would indicate. Mathematics has probably been responsible for the inflexibility in our understanding of the nature and function of these curves. Changing one parameter of a conic curve even the slightest amount can invalidate pages of difficult computation, so we avoid even thinking about what happens if we move a focus slightly.

TELESCOPES

If all parabolas had to operate exactly on axis and at the focal point, astronomical telescopes would be limited to peering at one star at a time—always precisely at the focus. We know that this is not the case. Reflecting telescopes can focus on a pattern of stars in the area around the primary focus. Stars to the right of the telescope axis will be focused correspondingly to the left of the axis. In practice, all reflectors based on conic curves can be operated a significant distance around the focus. Some airborne radar reflectors change the angle of the radar beam by maintaining the feed horn on a fixed plane while tilting the reflector. By tilting the reflector 5°, the beam is deflected 10°, supporting the rule that the angle of reflection is equal to the angle of incidence.

SOUND APPLICATIONS

Sound reflected from a parabolic reflector also forms an image of sorts. Conventionally, a single microphone is placed at the focus of a parabolic reflector aimed at the sound source. One supplier has come out with an assembly of two parabolic reflectors, each with a microphone at its focus, so that stereo recordings may be made. This is not only awkward and unnecessary, but it does not always give a true stereo output. To obtain stereo sound from a parabolic reflector, it is necessary only to mount two microphones in one reflector—one on either side of the focus. Figure 8-1 shows A parabolic reflector designed specifically for stereo reception. Note that the dish is considerably deeper than those used in conventional radar systems. Radar reflectors are frequently used for sound pickup simply because they are available in military surplus stores.

FIGURE 8-1

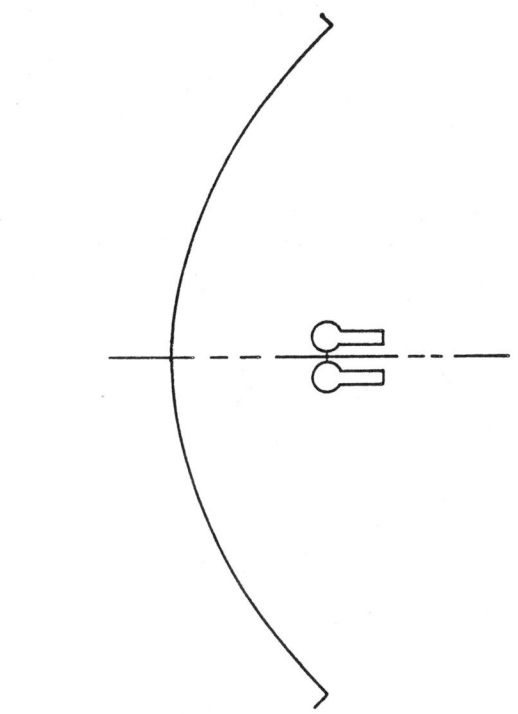

Parabolic microphones should be made with as large a diameter as possible, since the directional properties are an inverse function of the wavelength of sound. To be effective, the diameter should be several times the wavelength of the lowest frequency involved. Considering that the wavelength of middle C is about 4 feet, even medium frequencies require an impractically large reflector. Actually, the articulation sounds that carry speech information fall in the frequency range from about 800 to 4000 Hz. It is obvious that these highly directional microphones are not suitable for recording music, but they are quite effective when used to gather information. Bird watchers find that a 2- or 3-foot parabolic microphone operates quite well, since the frequencies involved are usually above 800 Hz.

VARIATIONS

By taking some liberties with the focus of a reflector, it is possible to make one curve perform the function of another. Figure 8-2(a) shows a parabola with a tangent drawn at its focal width. The tangent at this point must always be 45° since this is the only angle capable of reflecting a ray parallel to the axis at that point. Figure 8-2(b) shows the same parabola with the energy source moved a short distance forward of the focus, so that a ray leaving perpendicular to the axis hits a point on the curve that has a tangent less than 45° (such as an ellipse at its focal width) and is reflected toward a point on the axis. Although only one ray is traced in the drawing, the same principle applies to all rays, and we have an elliptical reflector for all practical purposes.

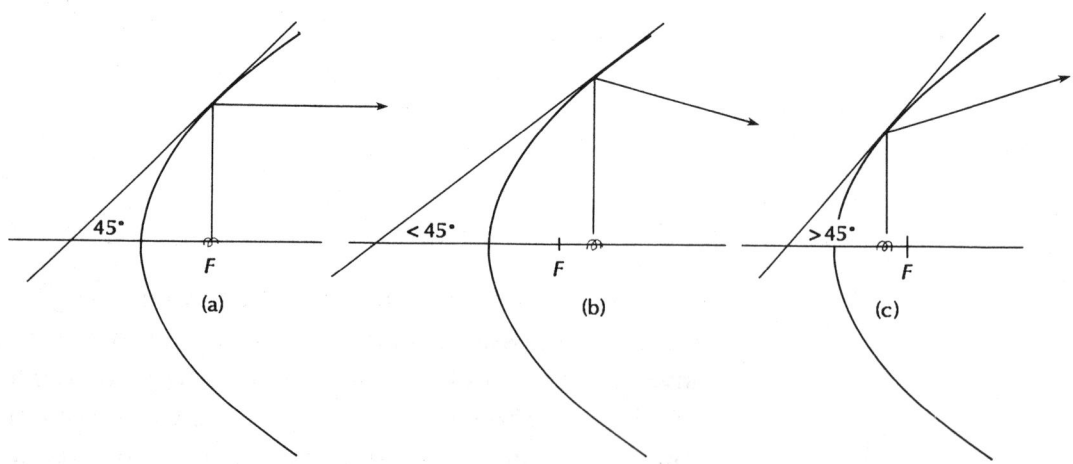

FIGURE 8-2

Figure 8-2(c) shows the energy source moved in toward the vertex. A ray leaving this point perpendicular to the axis now hits the curve at a point where the tangent is greater than 45°

(such as a hyperbola at its focal width), causing the ray to be reflected away from the axis. All rays are reflected in a divergent pattern. For practical purposes we now have a hyperbola.

The shallower the reflector, the easier it is to use it in modes other than those for which it was originally intended. For example, a parabolic radar dish whose focal length divided by its diameter (f/d) equals 0.4 or more may be used as an ellipsoid or hyperboloid within reasonable tolerance and may function in these modes.

Mathematical purists will shake their heads at this, but it is no worse than focusing a 50-mm camera lens at 60 mm in order to take a picture of an object close to the camera. Figure 8-3 shows why it is wise to use shallow reflectors when using a paraboloid for an ellipse or hyperboloid. All the curves are identical at the point where they cross the axis and remain quite similar in the immediate vicinity of the axis, but take on their own characteristics before reaching focal width.

FIGURE 8-3

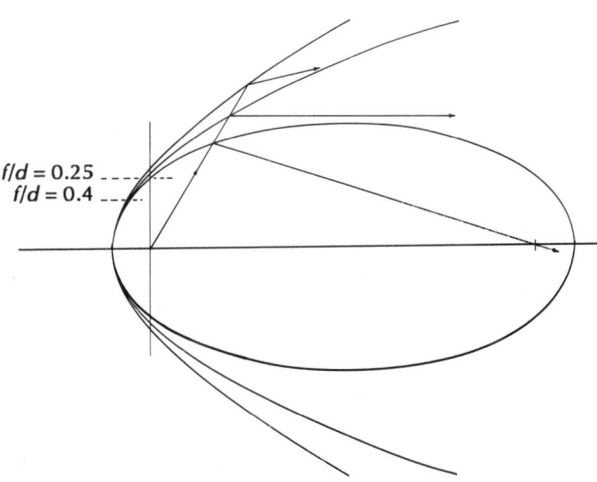

Since all energy sources (light, microwave, and so on) are of finite size, we are always compromising when we put something at the focus. For example, all lamp filaments have length. What point on the filament will have the honor of being at the true focus? At best, we project an image of the filament. The parts of the filament that are to the right of the focus will project to the left of the axis and vice versa. This can be used to our advantage. A deep ellipsoidal reflector with a filament placed on one axis can assemble a reflected image of the filament, complete with much of its heat. This can be projected through a glass wall into a vacuum chamber, if desired, and can heat an object by radiation without having to run wires into the chamber.

An electric Bunsen burner has been patented that uses the image of a filament as a heat source, as shown in Figure 8-4. The filament is placed in line with the axis and starts at the near focus of the ellipsoid. An inverted image of the filament is projected to the other focus outside the glass envelope.

FIGURE 8-4

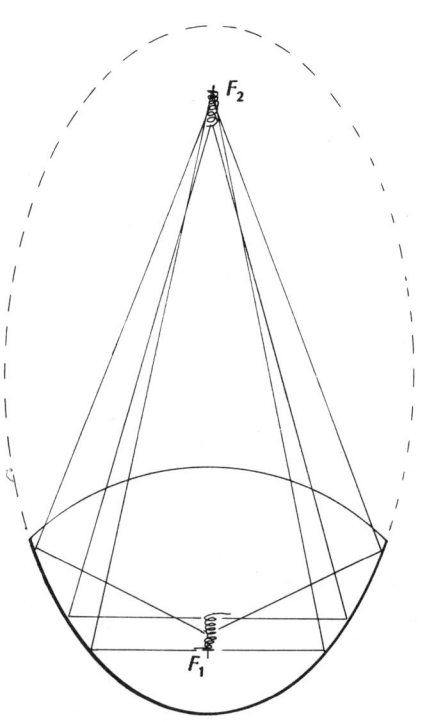

One major lamp manufacturer sells a 75-watt lamp in an ellipsoidal glass envelope. It is designed for use in recessed ceiling fixtures and is more efficient than the usual dispersive-type reflectors, since the light passes through a focus outside the light fixture and disperses after passing through the second focus. Incredibly, the filament is placed across the major axis instead of along the axis.

ACOUSTICAL APPLICATIONS

Efforts to utilize conic curves in theaters and churches have had comparatively little success. Elliptically domed theaters have been acoustical disasters because theaters require an even dispersion of sound instead of focal points for sound dispersion. Theaters so constructed have excessively live spots, usually in the balcony. A short distance from these spots, it is nearly impossible to hear. Also, since both foci of an ellipse work equally well in both directions, noise generated in the balcony can be very distracting at certain spots on stage.

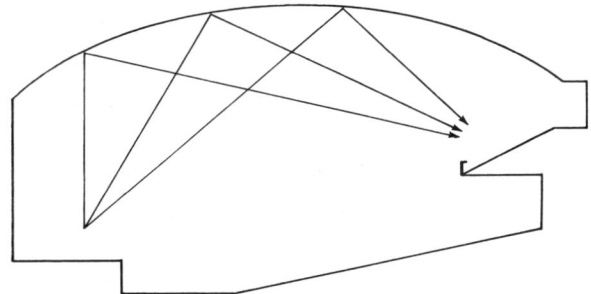

FIGURE 8-5

Theater with domed ceiling.

Band shells frequently approximate a parabolic shape, but again, if a true paraboloid were to be used, who would have the honor of sitting at the focus? They are actually designed so that sound generated anywhere inside is reflected forward to the audience with a minimum of vertical dispersion.

Parabolic pulpits, known as *sounding boards* (not to be confused with sound boards, which are used in pianos, harpsichords, and other string instruments), were fairly popular in churches a century ago. They consisted of large parabolic reflectors placed behind the pulpit; these reflectors had to be quite large and deep. They were abandoned when electronic amplification became practical. One obvious disadvantage of the parabolic reflector used for this purpose is its ability to gather sound as well as to project it. A dropped hymnal or someone snoring in the congregation would be picked up clearly by the preacher.

A method of directing sound in an auditorium has been devised that avoids the problem of point foci inherent in parabolic reflectors. Although it does not involve conic curves, it does involve geometry and is worthy of mention. It consists of an array of corner-cube reflectors in panels located behind the performers (Figure 8-6). (A corner-cube reflector is an open pyramid consisting of three right isosceles triangles joined at their right angles. The three hypotenuses form an open equilateral triangle.) Corner-cube reflectors have been used in radar and optics for many years but have only recently been tried in acoustics. They have the ability to return energy to its source as long as the source is within the quadrant of the corner cube. A number of these reflector panels return sound to the area of the performer. Since there are no focal points involved, it functions for all the performers equally well from anywhere on stage. There are two advantages in this system. The performers can hear themselves better than when conventional flat panels are employed, and the audience gets the sound, both original and reflected, coming from the area of each performer. This preserves a stereo effect, which is frequently lost in concert halls.

Dinkelspiel Auditorium at Stanford University in California is equipped with an array of 32 acoustical corner-cube panels. These have been in service for several years and have proved to be quite successful.

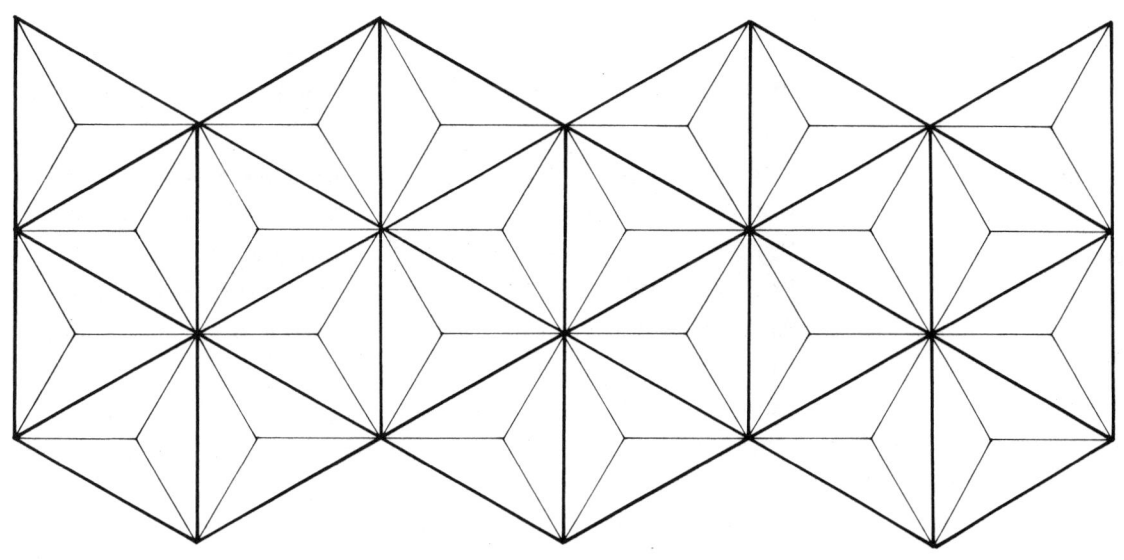

FIGURE 8-6

An array of 24 corner-cube acoustical reflectors.

CHAPTER NINE

UNUSUAL PROPERTIES OF CONES AND CONIC CURVES

The ancient Greeks believed that conic curves were of divine origin, since they had so many unusual properties. At that time they had barely scratched the surface. Nothing was known of their reflective abilities at the time of Euclid and Applonius, and elliptical planetary orbits were not identified for nearly two millennia. Pursuit of the geometric properties of these curves may not show them to be divine, but their natural occurrence and the relationships between the foci and axes can be a little spooky.

Ellipses share many attributes of circles, which are, of course, only special kinds of ellipses with foci that are coincident at the center. We have seen how ellipses and circles share a formula for area. In an ellipse, $A = \pi ab$, but when a and b are equal, the formula becomes $A = \pi r^2$. Later in this chapter we shall see that elliptical gears with axles at the foci will mesh as well as circular gears, although their performance is quite different.

AN INTERESTING PROOF

It is not the purpose of this chapter to bring out the many geometric problems and solutions that have traditionally filled geometry books of the past centuries. One, however, will be shown because it illustrates more than any other the uniqueness of these curves.

If we were to insert a sphere in a cone and pass a plane through the cone in such a manner that it just touched the sphere, the point of contact with the sphere would be at one focus of the ellipse formed by the cutting of the cone. This explains Figure 1-1, in which a ball rests on one focus of its elliptical shadow.

Now, if we insert another sphere in the cone so that it is tangent both to the cone and the ellipse, it will touch the ellipse at the other focus.

FIGURE 9-1

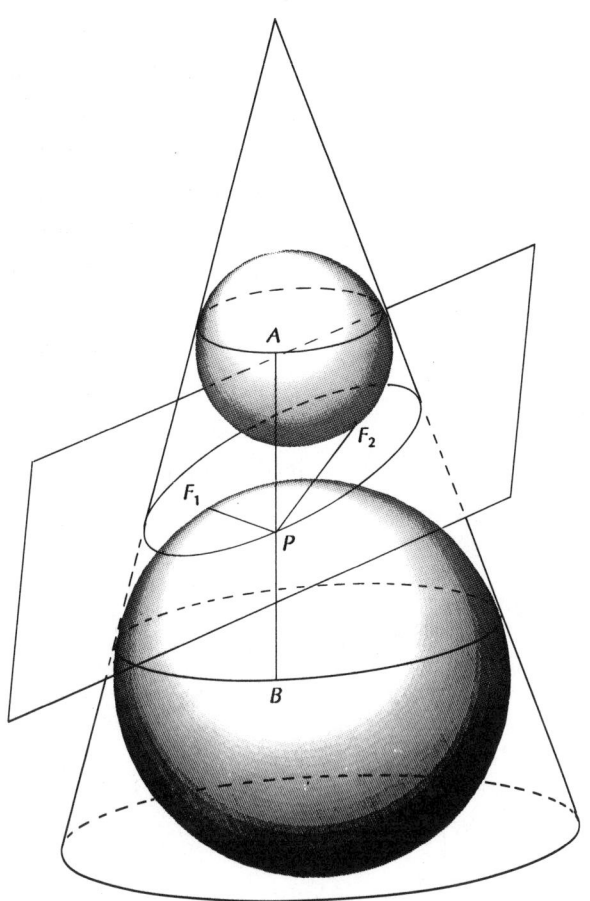

The proof for this is surprisingly simple and was discovered by Germinal Dandelin, a nineteenth-century Belgian mathematician. Figure 9-1 shows two spheres in a cone cut by a plane that is tangent to both spheres. Select any point P on the perimeter of the ellipse and draw a line through this point toward the tip of the cone and extending to the spheres at points A and B. The lines PB and PF_1 are the same length, since they are both tangents to the large sphere from point P. The same logic applies to lines PA and P_2.

Since $PB + PA = PF_1 + PF_2$, we fulfill the definition of an ellipse as the locus of a point whose distance to two fixed points (F_1 and F_2) is the length AB, which is also equal to the major axis of the ellipse.

Note that PF_1 and PB do not appear to be the same length in the figure even though we have proved them to be equal. This is due to the representation of a three-dimensional object in only two dimensions. The line PF_1 appears shorter because it is going away from point. This applies to lines PF_2 and PA, which are also equal to each other.

This geometric approach is general. It applies equally well to cylinders (which are portions of cones having 0° apex angles). The rule also applies to parabolas and hyperbolas, as shown in Figures 9-2(a), (b), and (c).

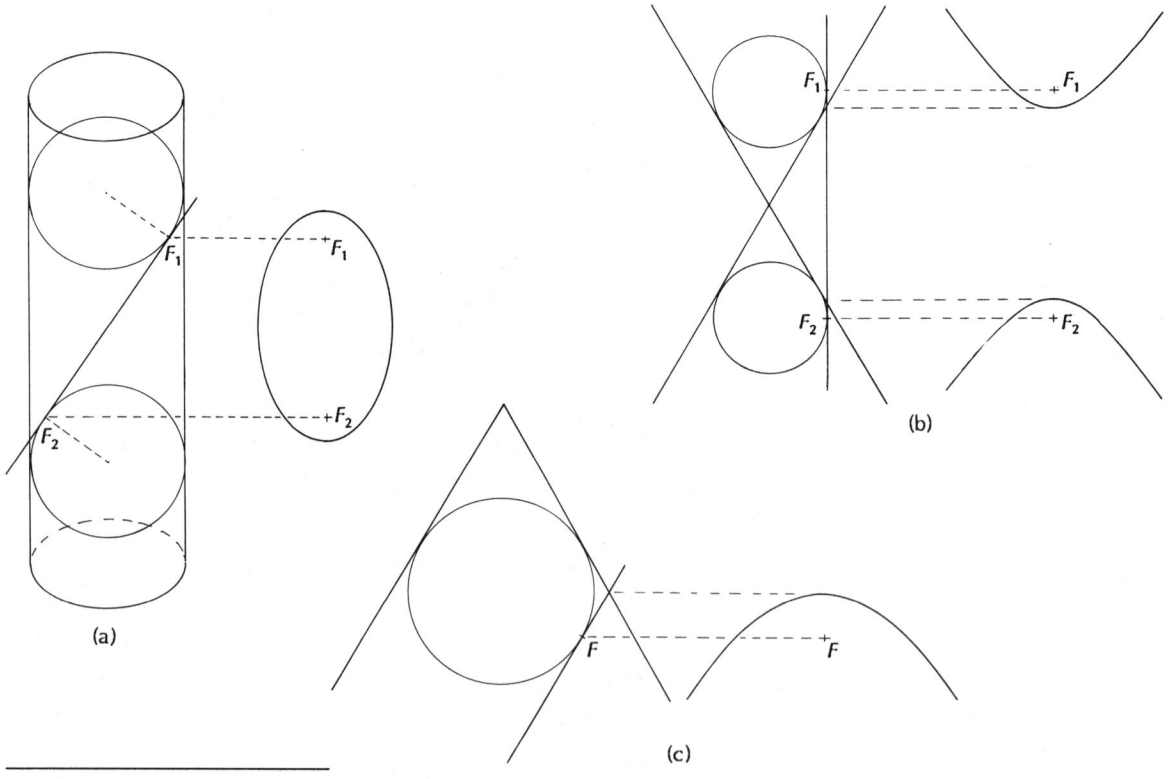

FIGURE 9-2

CONES

Cones may be thought of as being generated by a rigid stick pivoted at one point in a bearing, allowing the stick to swivel around a circle beneath it. (Strictly speaking, an ellipse or any other regular or irregular figure may be used in such a manner that a section sliced parallel to the base will yield a similar figure of different size. Even more strictly speaking, the pivot point does not have to be directly above the center of the circle. Such a cone is called *scalene,* or *oblique.* Discussion of exactly what happens when these cones are cut up would cloud the issue, so they are only mentioned in passing. Only right circular cones are considered here. If the rigid stick passes through a pivot point at its middle and passes around a circle at one end, another cone is generated point-to-point, or apex-to-apex, with the one underneath. This is important because hyperbolas depend upon two cones for their generation by a plane intersecting the cones. Figure 0-2 in the Introduction illustrates this.

Witch hats and dunce caps were traditionally conical. There may not be any greater significance to this than the fact that cones are easily made from any available material (such as paper) that can be rolled on one plane but not on two planes at the same time. Dunce caps are easier to make than derbies.

If a right circular cone is cut off (truncated) parallel to its base, the bottom part is called a *frustum.* If the cone is cut off obliquely to its base, the result is called an *ungula* and resembles a hoof; in fact, it gets it name from the Latin word for hoof.

Cones have been useful since the invention of the funnel, which has the ability to channel liquids or granules poured from one or more sources into the large end and to concentrate them at the small end.

Megaphones perform the opposite function of funnels. They take sound concentrated at a small source and direct it out the large end. This explanation is really too simple, since several factors enter into the operation of a megaphone. Since the angular spread of sound from its source is inversely proportional to the wavelength of sound and directly proportional to the diameter of the output end, the higher frequencies become quite directional. The same rules governing the spread of sound as a function of diameter and wavelength apply here as described in the discussion of parabolic microphones in Chapter 8. The higher the frequency and the larger the diameter of the large end of the megaphone, the more effective it is in directing sound energy. The higher frequencies are the frequencies of articulation (hisses, clicks, and so on, which are above 1000 Hz) that carry the information of speech, so megaphones are very good at concentrating the available speech energy and directing it to the desired location. They are not, however, very good for directing the full range of musical sound.

Megaphones have the ability to couple high-impedance acoustical sources (vocal chords and speaker cones) to the relatively low impedance of free air and may be thought of as acoustical transformers. It was in this capacity that cones were used in the old acoustical phonographs, and they survive today in a slightly more sophisticated form as exponential horn speakers.

Not all energy is as easily gathered as sound. Light, for example, cannot be concentrated by gathering it in the large end of a cone. Since light is reflected from the walls of the cone, each reflection picks up the angle of the cone wall, directing it toward the other side and increasing its angle of incidence at each reflection, until it is turned around and headed out the way it came in (Figure 9-3).

FIGURE 9-3

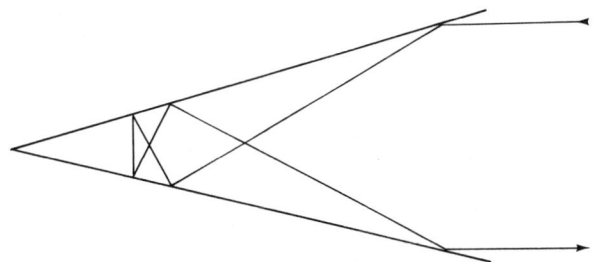

Cones (or portions of cones) roll around in a circle on a flat surface, the apex of the cone being the center of the circle (Figure 9-4). Some sea birds that nest on rock cliffs, such as the murre, lay highly conical eggs. If an egg rolls out of the nest onto a flat surface, it rolls in a circle and returns to the nest.

FIGURE 9-4

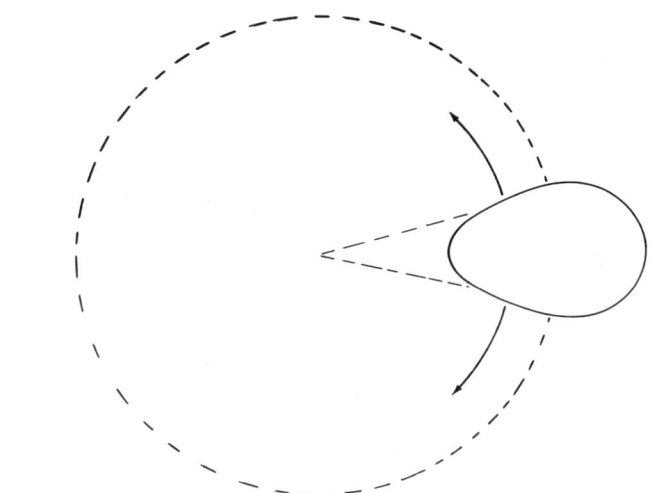

The shortest path around a cone is not always a conic section. If a spider wishes to walk around a cone and return to its starting point, covering the least possible distance, it will follow the path shown in Figure 9-5. The logic of this is apparent if we cut the cone open from the apex to the base (on the side opposite the starting point) and spread it out flat. It is obvious that the shortest distance around the cone from the point of origin back to the point of origin will be along straight lines perpendicular to the cut line. If the cone is equilateral or even more obtuse, the spider would have to climb straight up to the apex, turn around, and return by the same path.

FIGURE 9-5

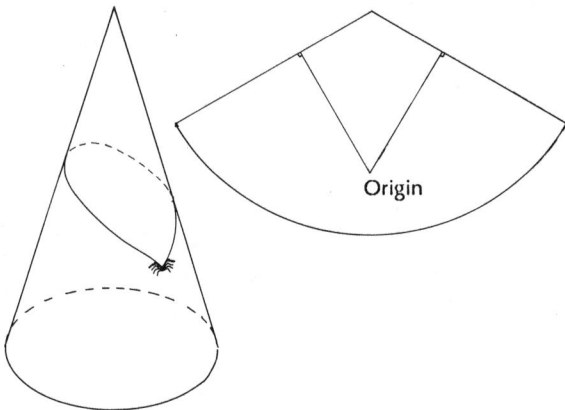

ELLIPSES

Ellipses have interesting properties other than reflecting energy that passes through (or originates at) one focus to the other focus. What happens if we miss the focus? Let us create an imaginary frictionless elliptical billiard table and roll a ball so that it passes between a focus and one end of the table. The ball will never pass between the foci but will develop an envelope of tangents to another ellipse having the same two foci as the billiard table (Figure 9-6).

FIGURE 9-6

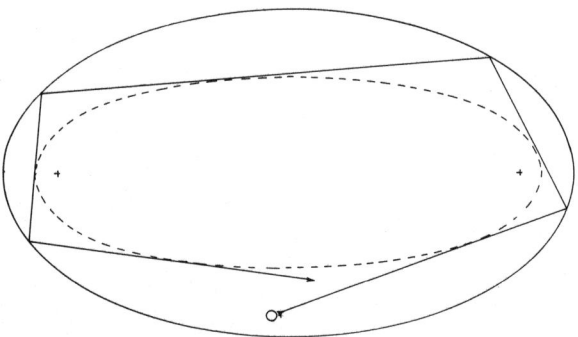

If the ball passes between the foci, as shown in Figure 9-7, it will develop an envelope of tangents to a pair of hyperbolas also sharing foci with the elliptical table.

FIGURE 9-7

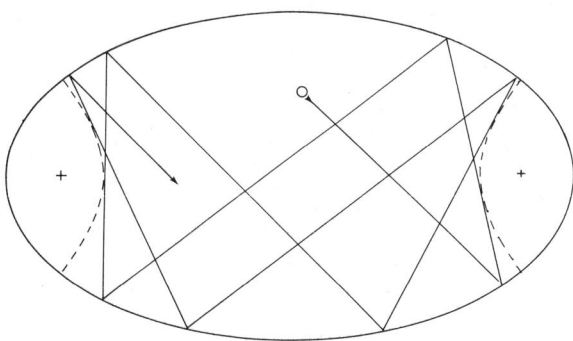

If the ball passes through one focus, we can expect it to pass through the opposite focus after each reflection. The surprising aspect of this is that with each reflection, the path of the ball becomes more nearly coincident with the major axis of the ellipse. For practical purposes, it lines up quite well with the major axis after a few passes (Figure 9-8).

FIGURE 9-8

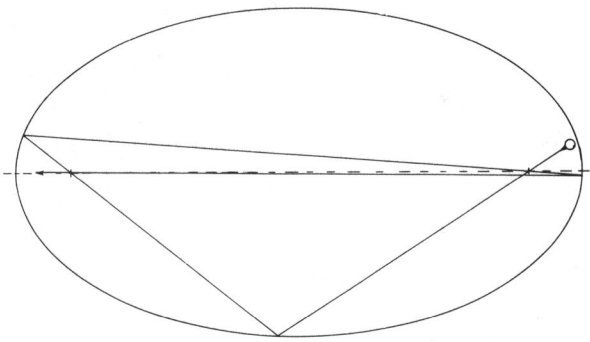

KEPLER'S LAWS

No discussion of ellipses would be complete without the mention of two laws formulated by Johannes Kepler in the early seventeenth century. His first law states that the planets move in elliptical orbits, with the sun at one focus. Figure 9-9 shows this principle but with the eccentricity greatly exaggerated for any of the planets in our solar system. Actually, all the planets have very nearly circular orbits, with the exception of Mercury and Pluto. Since the orbits of the other planets deviate so little from circular orbits, they cannot

FIGURE 9-9

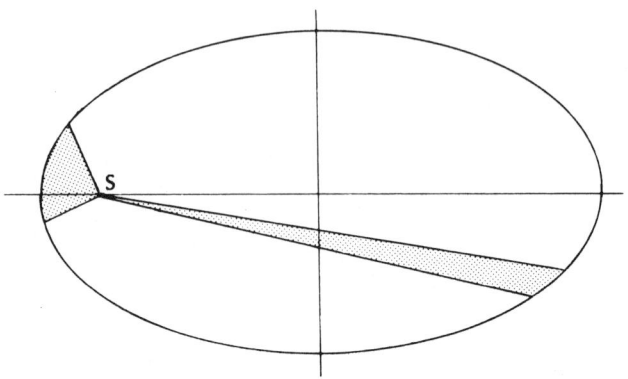

effectively be drawn to scale. Figure 9-10 shows a circle and the orbits of Earth (E), Mercury (M), and Halley's comet (H). For comparison, their major axes have been adjusted to be the same as the diameter of the circle. The orbit of Earth falls within the width of the ink line of the circle. The orbit of Mercury, considered by planetary standards to be quite eccentric, really deviates little from circular. Halley's comet, however, is quite eccentric, passing close to the sun and quickly passing out of the Solar System for another 76 years. The orbit of the comet Kohoutek is so eccentric that its minor axis would be less than the width of the major-axis line in this drawing.

FIGURE 9-10

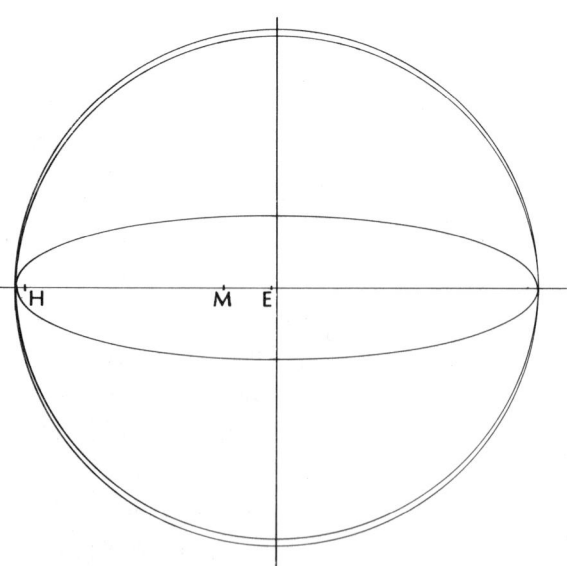

The shaded areas of Figure 9-9 are triangles of equal areas. According to Kepler's second law, a planet takes the same time to cover the arcs of the orbit determined by these equal-area sectors. The speed of the planet is greatest when it is nearest the sun and least when it is farthest from the sun. The planet literally "falls" toward the sun. Our planet is nearest the sun (perihelion) during winter in the northern hemisphere. After the winter solstice, the sun does not start rising earlier for about two weeks. The earth travels more than 1/365 of its orbit each day when it is closest to the sun, and the angle covered is correspondingly greater than it would be if the orbit were circular. Although this causes the sun to rise later than expected immediately after the winter solstice, it also makes the sun set later, so that the total amount of daylight increases even though the sun continues to rise later for a while.

There is an interesting sidelight to Kepler's second law. If, while a planet is orbiting the sun, the sun were to lose its gravity, the planet would proceed in a straight line. Would the law of equal area for equal time continue to be valid?

FIGURE 9-11

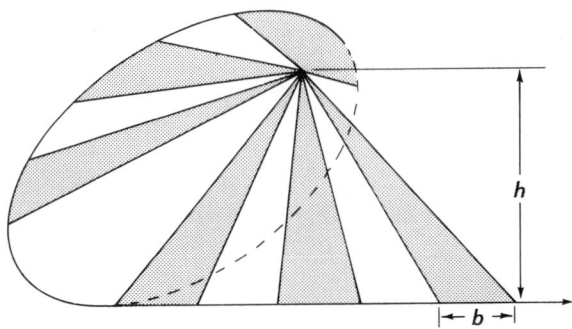

The answer is yes. Figure 9-11 shows that triangles would continue to have equal areas for equal times, since the bases of all the triangles would be the same (in linear motion, the speed is constant) and the height of all the triangles is equal. The area of a triangle is equal to half the base times the height. Since all the triangles have equal bases and heights, they must also have the same area. Actually, if the sun were to lose its gravity, there would be nothing special about the place where the sun's center of gravity had been. Any point not on the line described by the planet would be as good as any other point, since all the resulting triangles would have the same base and height.

This highlights the concept that there is little difference between a body moving in a straight line and one that is in orbit around a gravitational body as far as the moving body is concerned. With our present knowledge of gravity and orbital motion based on Newton and Einstein, it is easy for us to see the logic in Kepler's second law. Unfortunately, Kepler did not have our knowledge and arrived at his laws through years of tedious calculations and at least one correct assumption. He used the Ptolemaic concept of the equant, modified to fit the Copernican model of the Solar System in which the planets orbit around the sun. An *equant* was a point displaced from the center of an orbit equal to the displacement of the sun on the other side of the center (Figure 9-12).

FIGURE 9-12

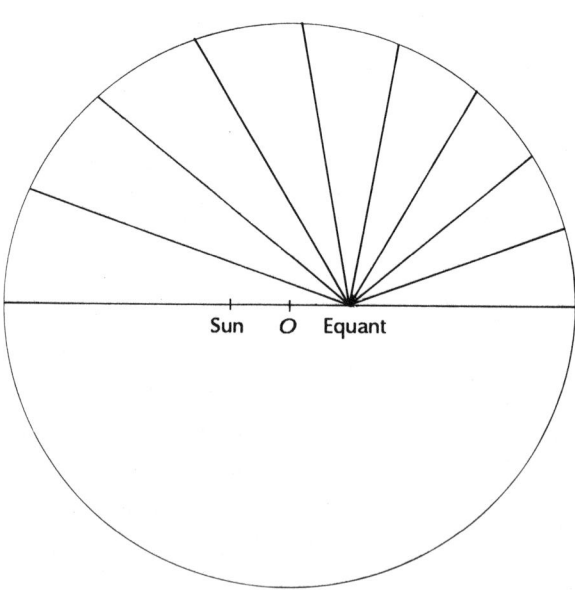

The equant was introduced by Ptolemy in an effort to explain the measured variations in orbital speeds of the planets and meant that the center of the orbit was an unoccupied point in space midway between the sun (or Earth in the Ptolemaic model) and the equant. When radial lines of equal angle are drawn from the equant, it can be seen that they intercept the circular orbit to form unequal arcs around the circle (Figure 9-13). If we connect lines from the sun to the intercept of these radials, we find that angles on one side of the circle are larger than angles on the opposite side. Near the line joining the sun, the center, and the equant, these angles will form sectors that have approximately equal areas. Kepler's belief that equal areas were swept in equal time by an orbiting planet came before its proof. The equant swept-area rule applied only when the sectors were very near these points but were not valid for any other parts of a circular orbit.

FIGURE 9-13

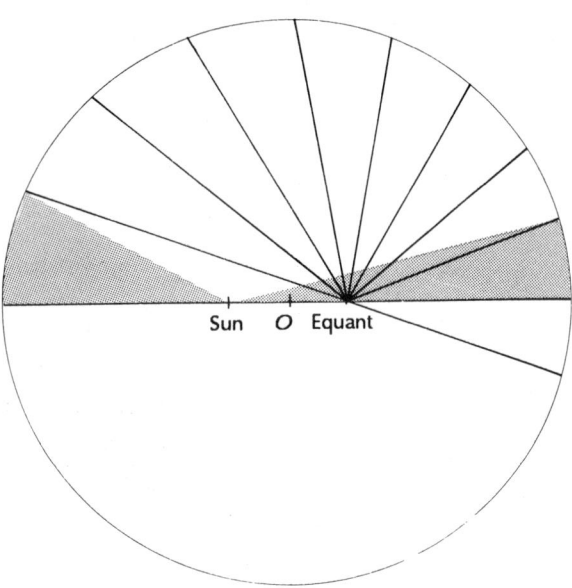

The position of the equant and the sun might make us think immediately that they were the two foci of an ellipse. Unfortunately, conic sections were not well known in Europe at that time and Kepler tried a number of "ovals" to see if they

79 | CHAPTER NINE: UNUSUAL PROPERTIES

would fit his model. When he found that an elliptical orbit with the sun at one focus conformed to his calculations, he was ecstatic. (A tracing of Kepler's proof that equal areas were swept in equal time in an elliptical orbit can be seen on page iv). Kepler also showed that ellipses (and other conic sections) are no less divine than circles. Figure 9-14 shows the modification from circular to elliptical orbit necessary for the law of equal swept areas.

FIGURE 9-14

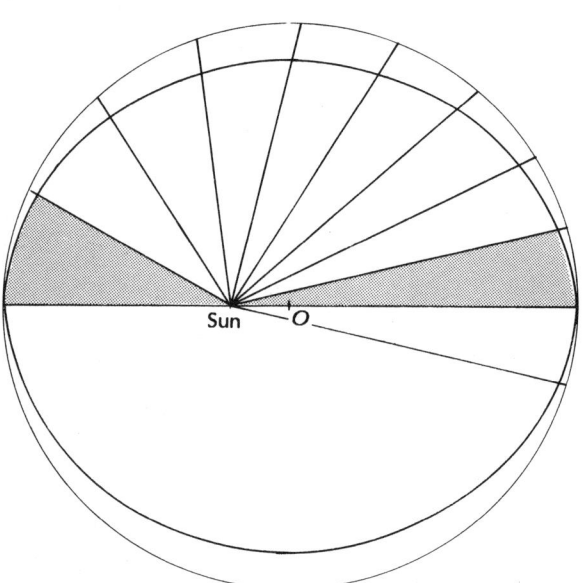

Nothing impedes the advancement of knowledge as much as the belief that we already have that knowledge. For example, if you "know" that Earth is flat, you will not go looking for a round Earth. The dead weight of established errors of the past made progress difficult. Aristotle had established that orbits of planets had to be circular. This was based on a quasi-religious belief that celestial objects, being of divine origin, had to move in the most perfect of all curves—that is, circles. It became obvious to Ptolemy in A.D. 150 that circles did not explain orbital motion, so his model of the Solar System had planets making smaller circles while executing their large orbital circles. These were known as *epicycles* and explained the periodic retrograde motion of the planets, which appeared to stop and reverse their motion briefly, and then continue. (The name *planet* comes from the Greek word for wanderer.) We would like to think that Copernicus solved all this nonsense by putting the sun in the center of the circular orbits, but his model of the Solar System turned out to be more complicated than Ptolemy's. Ptolemy's model had 40 epicycles and Copernicus's had 48.

Try to imagine the nearly insurmountable obstacles that faced Kepler at the time he discovered the laws of planetary motion. During Kepler's lifetime, Giardano Bruno was burned at the stake by the Inquisition for believing (and teaching) that the sun was at the center of the Solar System, and Galileo spent his last years under house arrest for the same "crime." Students in some European universities could be fined for expressing thoughts that were contrary to the teachings of Aristotle. Besides the official suppression of science, superstition and misconceptions that had been accepted during the Middle Ages formed an equally difficult obstacle to the advancement of knowledge. For example, Kepler believed that planets had souls, and he spent much of his time trying to fit the planets into a system of geometric solids. His only marketable skill was that of an astrologer.

It is little wonder that it took so long to discover the relatively simple structure of the Solar System when you consider that Earth, from which all our observations are made, is in an orbit around the sun with two planets inside our orbit and the remainder of the planets outside. There is no fixed observation point from which to obtain a clear picture of the Solar System.

About fifty years after Kepler formulated his laws of planetary motion, Isaac Newton perfected the mathematics for their proof. Kepler believed that an invisible force from the sun drove the planets in their orbits and that the planets would stop if this force ceased. Newton's laws of motion and gravitation gave us an understanding of what the planets had been doing all along and provided the mathematics required to determine the orbits of spacecraft orbiting Earth and the excursions to the outer planets of the Solar System.

ORBITAL VELOCITIES

While on the subject of elliptical orbits, there are two interesting items concerning orbital velocities. Figure 9-15 shows an elliptical orbit with the sun at one focus in a circle with a radius equal to the major axis (2*a*) of the ellipse and centered at the focus. An object in orbit will have, at any point, the same velocity as an object falling from the circumference of the circle as the falling object crosses the ellipse on its way to the sun.

FIGURE 9-15

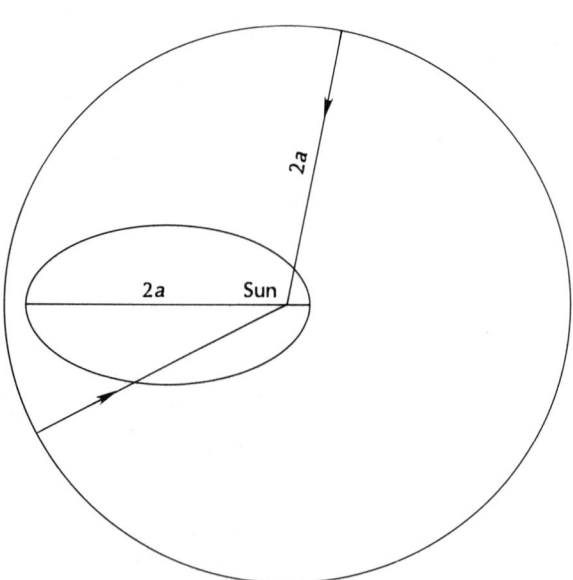

The length of the major axis of an elliptical orbit determines the orbital period. All bodies (satellites, planets, comets, and so on) in orbit around the same center of gravity and having equal major axes will have equal periods. At any point at which the orbits cross each other, the velocities of the orbiting bodies will be the same. Obviously, the more circular the orbit, the greater the distance traveled by the orbiting body, since the circumference of an ellipse is less than that of a circle of the same diameter as the major axis of the ellipse. The length of time to complete the orbit will be the same.

ELLIPTICAL GEARS

Elliptical gears may be paired by rotating them about their centers or their foci. Figure 9-16(a) shows a pair of gears with axles at their centers. The ratio of the speeds of rotation will change continuously as the ratio of the major and minor axes. If the driver gear maintains a constant speed, the other gear will alternately speed up and slow down twice during each revolution.

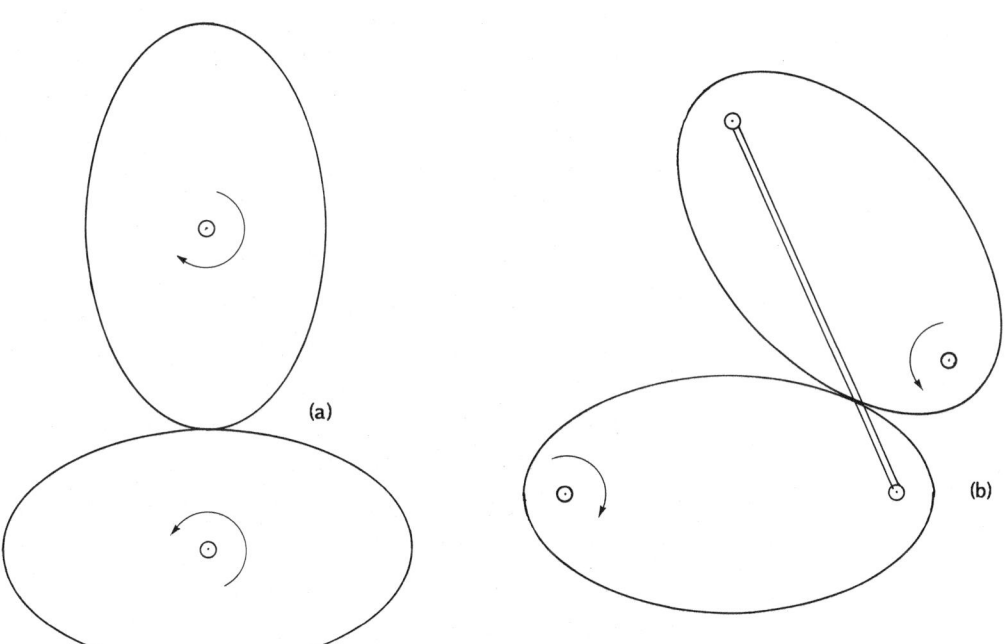

FIGURE 9-16

Figure 9-16(b) shows two elliptical gears with axles on the foci. The driver gear turns at a constant speed, and the driven gear changes its speed continuously during each revolution.

In the illustration the change amounts to approximately 10:1 at one point and 1:10 half a turn later. This means that the speed of the second gear varies in a ratio of 100:1! As the gears turn, the distance between the foci that are not on axles remains constant and is the same distance as the major axis of the ellipses. In practice, a linkage arm is usually connected between the foci that are not used as axes of rotation. This is done to keep the gears in mesh.

If you wish to experiment with elliptical gears, you may make them by laying out a suitable-size ellipse (templates are readily available at drafting supply stores) on a piece of plywood or hardboard and cutting it out with a sabersaw. Gear teeth can be added by gluing a toothed belt on the edge of the ellipses. Various printer and timing belts are available from electronic surplus stores. If you are lucky you can lay out the ellipses to be exactly the same size as the available belts. Actually, you will probably have to cut the belt to the proper length and glue it to the edge of the ellipse with contact cement. Some spring tension will probably be required between the axles to keep the gears meshed.

SKETCHING CONICS FROM A FORMULA

To sketch an ellipse or hyperbola from the information given in a formula, we need to construct a rectangle that has sides equal to the major and minor axes (the dimensions $2a$ and $2b$). Once this is done, we may sketch an ellipse inside the rectangle, making sure that the only points of tangency with the rectangle are the intercepts of the major and minor axes (Figure 9-17). To establish the foci, all that is necessary is to swing a compass set to the dimension a from either of the minor-axis intercepts.

FIGURE 9-17

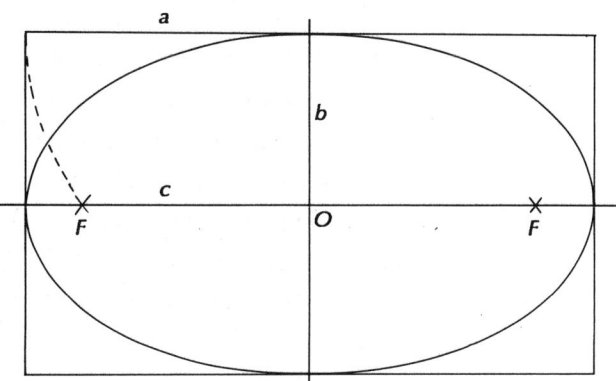

A circle, which is a special case of an ellipse, fits nicely inside a square. All the rules of the ellipse apply to the circle, but we find that both foci are coincident with the center, since the major and minor axes are equal.

To sketch a hyperbola from a given formula, we start with a rectangle whose sides are 2*a* and 2*b,* just as we did with the ellipse. This time we draw extended diagonals across the rectangle. These diagonals are the asymptotes of the hyperbola, which is drawn outside the rectangle. When sketching the hyperbola, we must make sure that the vertex is tangent to one of the sides of the rectangle and that the curve never touches the asymptotes (Figure 9-18).

FIGURE 9-18

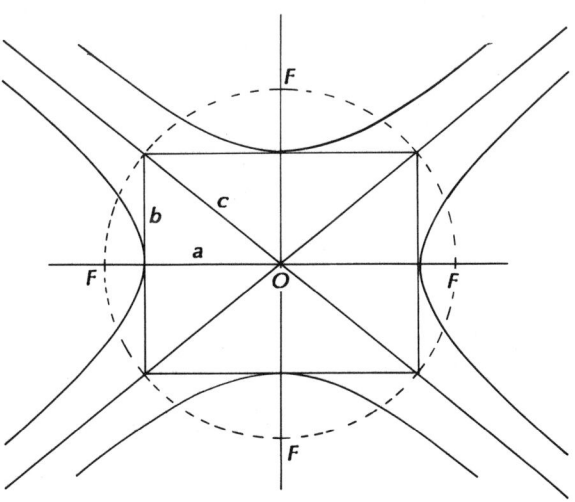

Finding the foci of hyperbolas appears to be a little more tricky than finding the foci of ellipses, since there is no minor axis crossing the curve; therefore, there is no place on which to put a compass point. To add to the confusion, the dimension c is longer than a. This is resolved in Figure 9-18, which shows that the dimension c is half the diagonal of the rectangle. To find the foci, we simply swing a compass set to this dimension from the point O at the center of the rectangle.

So far, we have generated ellipses starting with the major axis and either the minor axis or the location of the foci. Let us pretend that we have found an ellipse that has none of these lines or points and that we wish to find its vital dimensions. As shown in Figure 9-19 we draw two parallel lines, AB and CD, through the ellipse. The exact angle and spacing of these lines is not critical, although they should have reasonable distance between them, but the lines must be parallel. We bisect these two lines and connect the midpoints with a line EG. Bisecting this line between the points of intersection with the perimeter of the ellipse will give the exact center (O) of the ellipse. Using a compass, we draw a circle large enough to intersect the ellipse at four points and draw the rectangle $HIJK$.

FIGURE 9-19

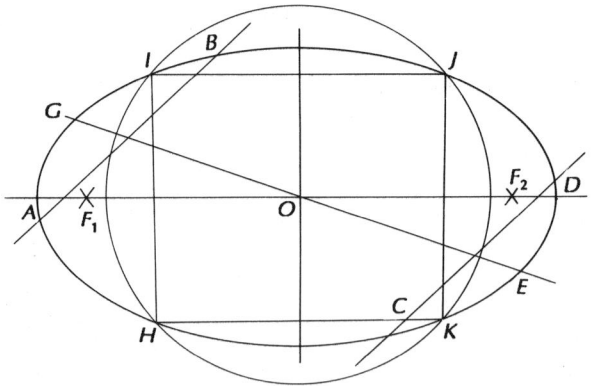

We now bisect the two vertical lines HI and JK and the horizontal lines IJ and HK. When we draw a line through the midpoints of HI and JK we have found the major axis.

Likewise, we connect the midpoints of *IJ* and *HK* to find the minor axis. To find the foci, we use a compass to take the distance from the center of the ellipse to the major-axis intercept and swing it from the minor-axis intercept. The points where it crosses the major axis are the foci.

It would be interesting to determine the major and minor axes, foci, and eccentricity of the Oval (elliptical) Office of the President of the United States. The thought of such an activity has probably never crossed the mind of any president since Thomas Jefferson.

Ellipses are frequently used as logos for corporations. Ford, for example, uses an ellipse with an eccentricity of 0.915.

We have shown how to find the axes and foci of ellipses, so that if we are confronted with an old Esso sign or even the Oval Office, we can determine its vital dimensions. Ellipses are closed (finite) figures and are usually complete, so there is less of a problem working with them than with parabolas and hyperbolas, which, no matter how large, can never be complete finite entities.

Parabolas have one advantage over the other two conic curves, since we know that no matter what the size, the eccentricity will be unity and the second focus will be at infinity. This is very important because it ties down one of the dependent variables and makes it easier to determine its focus, the only other variable needed to determine the curve. As stated in Chapter 3, the only real variable in parabolas is size. This same condition exists for circles.

Let us suppose that we have purchased a radar dish and have no idea of its focal length. Usually, but not always, its focal length is 40% of its diameter. This is fairly standard for radar dishes. To determine its focus exactly by geometry, lay a straightedge across the dish and measure its depth to the vertex. Using a suitable scale (if the reflector is too large to draw full size), draw an axis line and a perpendicular line that is tangent to the vertex of the parabola.

FIGURE 9-20

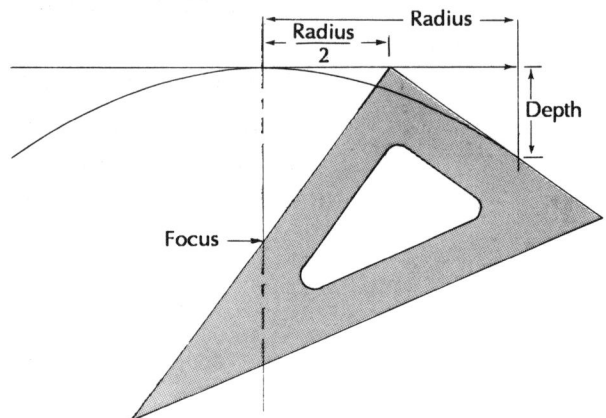

(It is necessary only to draw half of the curve, since it is a symmetrical figure of rotation.) For our purposes, let us assume that the reflector is 4 feet in diameter and has a depth of 10 inches. Scale out the appropriate points and place one arm of a right angle passing through the point known to be the edge of the dish 10 units from the line that is tangent to the vertex. The right angle must occur on the midpoint on the tangent line. The other arm of the right angle will cross the axis at the focus, as shown in Figure 9-20.

This may be a case in which algebra is easier than geometry. To find the focal length of this reflector, simply divide the square of the radius by four times the depth:

$$f = \frac{\text{radius}^2}{\text{depth} \times 4}$$

To create a weightless condition, NASA uses a specially equipped KC-135 aircraft to fly a parabolic path. The aircraft flies at full speed at a moderately low altitude and then pulls up into a parabolic trajectory. Power is reduced, and complete weightlessness is attained for nearly half a minute (Figure 9-21). This situation is important in the training of astronauts and in carrying out scientific experiments that can be done only in a weightless environment at a cost significantly less than an actual space flight. One of the interesting things found during one of these flights is that fires will not sustain themselves if there is no gravity. The hot combustion products cannot rise (there is no "up") to bring in new oxygen, and the flame extinguishes itself.

FIGURE 9-21

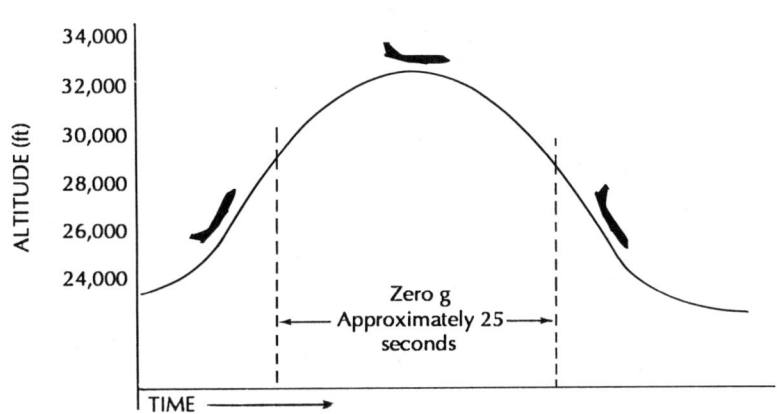

CONCLUSION

Conic sections are a part of our lives. Our fountains describe parabolas, our lamps cast hyperbolas against the walls of our homes and offices, and spotlights cast ellipses upon a stage. Even in our language we have *ellipsis* (the omission of a word necessary for complete grammatical construction, but understood in the context), *hyperbole* (an obvious exaggeration not meant to be taken literally), and speaking in *parables* (stories with a moral or religious lesson). Conics have been with us since the beginning and will be here long after humanity has departed.

APPENDIX A

NONPARABOLIC REFLECTORS FOR ANTENNAS

Sometimes it is desirable to make a reflective antenna that radiates and receives with a beamwidth greater than possible with a parabolic reflector. In this case either a hyperboloid or an ellipsoid can be used. The hyperboloid would operate very much the same as a paraboloid, but it would be more shallow than a paraboloid of the same focal length. An ellipsoid would be deeper than the equivalent paraboloid and would bring the energy to a focus some distance from the reflector, after which it would continue in a divergent pattern (Figure A-1).

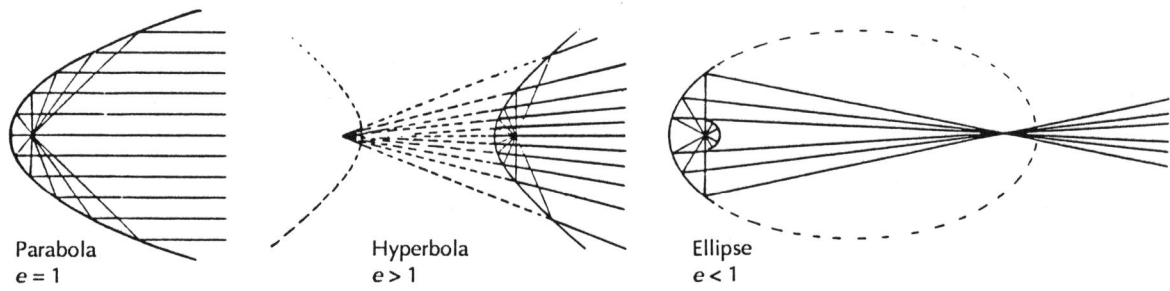

FIGURE A-1

Since the figures involved are so close to parabolic, the dimensions a, b, and c are usually too long to be laid out conveniently on paper by the conventional geometric construction methods demonstrated earlier in this book. It is easier in this case to use analytical geometry in designing the figures. The conventional nomenclature is given in Figure A-2.

It is more difficult to design hyperbolic and elliptical reflectors than parabolic reflectors because for the parabola the focal length will completely determine the proportions of the reflector, and parallel beams will be reflected. For elliptical and hyperbolic reflectors, the focal length ($a - c$ for the ellipse, and $c - a$ for the hyperbola) must be considered along with the degree of beam divergence. The dimensions a, b, and c, referenced from a point O, must all be considered.

FIGURE A-2

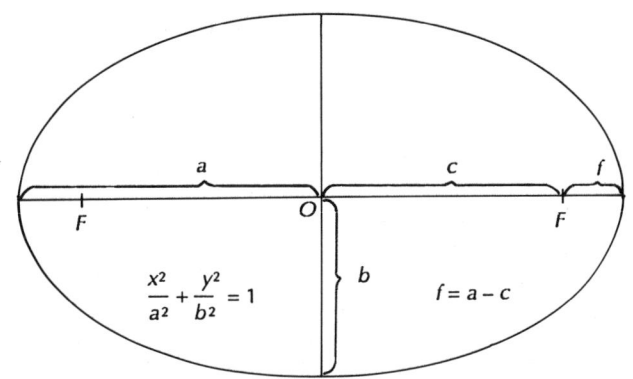

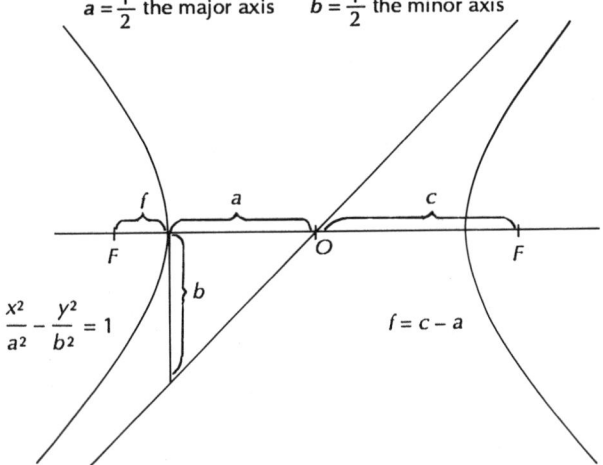

O = center of ellipse or pair of hyperbolas c = distance from O to focus

$a = \frac{1}{2}$ the major axis $b = \frac{1}{2}$ the minor axis

We can usually assume that the desired beam divergence will be small, seldom more than 10°. This means that the desired curve will be very close to parabolic. Since the parabola is the conic curve with an eccentricity of 1.00, it may be safely assumed that the required hyperbola would have an eccentricity only slightly greater than unity, or the required ellipse would have an eccentricity slightly less than unity.

The values of a, b, and c are not obvious when the only given values are the distance from the focus to the vertex and the eccentricity. However, this is sufficient information to calculate these values.

To find c and a:

Eccentricity: $$e = \frac{c}{a}$$

$$c = ea$$

Focal length: $$c - a = f$$

Substituting, $$ea - a = f$$

$$e - 1 = \frac{f}{a}$$

Therefore, $$a = \frac{f}{e - 1}$$

To find b: $$b^2 = c^2 - a^2$$

Substituting, $$b^2 = (ea)^2 - \left(\frac{f}{e-1}\right)^2$$

The example to be illustrated is the design of a 30-inch hyperbolic reflector with an eccentricity of 1.09 and a distance of 12 inches from the vertex to the focal point. It is to have a 3° beamwidth. We know that the eccentricity is 1.09 and that $f = 12$. From this information we can derive a, b, and c.

$$a = \frac{f}{e - 1} = \frac{12}{0.09} = 133.33$$

$$c = a + f = 133.33 + 12 = 145.33$$

$$b^2 = c^2 - a^2 = 145.33^2 - 133.33^2$$

$$b = 57.827$$

With the values for a and b, it is now possible to calculate the coordinates of the curve using the formula $\frac{x^2}{a^2} - \frac{y^2}{b^2} = 1$.

The 3° beamwidth is a little more difficult to calculate. From a practical standpoint, it is probably best to make an educated guess, draw the curve, and measure the angle of tangency at the perimeter. It must be remembered that the divergence (or convergence, in the case of ellipses) is a function of the diameter and that the angle of reflection is equal to the angle of incidence. For a beamwidth of 3° it is necessary to change the angle at the perimeter 0.75° from that of a parabola. This increases the angle 1.5° after reflection on each side of the reflector, giving us the desired 3° beamwidth.

It is possible to use the same mathematics to calculate the coordinates of an ellipse with only minor modifications. Since the eccentricity of ellipses is less than 1.00, the value $(e-1)$ is negative, and b is derived from $b^2 = a^2 - c^2$. These changes in the formula are obvious and should present no problem in deriving the coordinates.

Elliptical reflectors can be used both for the concentration and for the dispersion of energy. Note that the energy converging on F_2 will arrive and leave at a much narrower angle than the energy leaving the source at F_1. This can be used to advantage in the compound reflector shown in Figure A-3.

FIGURE A-3

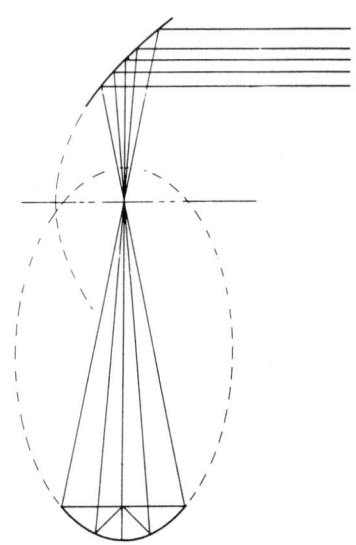

It is possible to concentrate a large amount of radio-frequency energy into a small volume with an ellipsoidal reflector. This property can be particularly useful, since it allows the second focus to be inside a vacuum chamber or in an area where it is not practical to have wiring or a waveguide, but where pure microwave energy is required.

Another proposed use for high-power concentration is in the clearing of mine fields for the military. An ellipsoidal reflector can be mounted in the front of an armored vehicle and fed by a powerful source of microwave energy. The second focus of the reflector is located approximately 50 feet in front of the vehicle. Several megawatts of energy can be concentrated, which is believed to be sufficient to detonate mines.

The source of energy at the focus may not always be represented as a point. For example, a "splashplate" feed is a ring source instead of a point. Energy leaving such a feed will not be from a point but rather a circle. To compensate for this, the axis of rotation of the parabola generating the paraboloid should be displaced by a distance equal to the radius of the ring source, as shown in Figure A-4.

FIGURE A-4

Parabola with displaced axis of rotation.

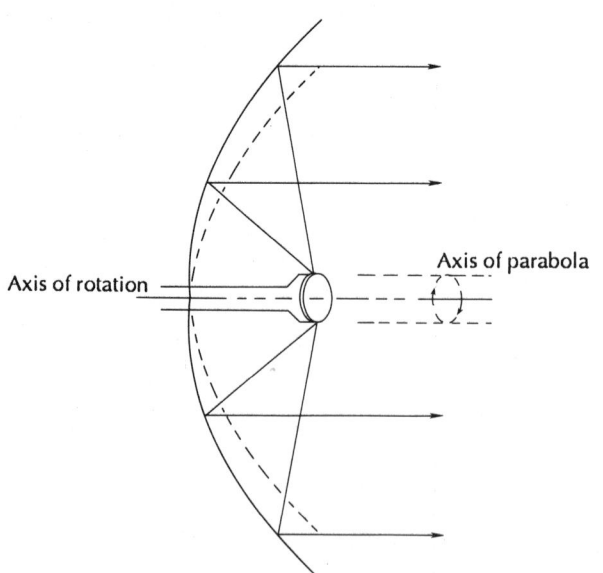

APPENDIX B

COMPUTER PROGRAM FOR GENERATING CONIC CURVES

This QBASIC program, written by Roy Bannon, will calculate the coordinates of the Conic Sections on an IBM computer.

```
      DEF SEG = 0
      MON% = PEEK(&H449)
      DEF SEG

MAINMENU:

      CLOSE
      OPEN "O", #1, "DATA1.$$$"
      CLS
      IF MON% <> 7 THEN
            SCREEN 9
            COLOR , 9
            LINE (10, 20)-(400, 250), , B
      END IF
      PRINT "          CO-ORDINATE CALCULATION PROGRAM"
      LOCATE 5, 3:  PRINT "( 0 )  EXIT TO DOS"
      LOCATE 7, 3:  PRINT "( 1 )  PARABOLA"
      LOCATE 9, 3:  PRINT "( 2 )  ELLIPSE"
      LOCATE 11, 3: PRINT "( 3 )  HYPERBOLA"
      LOCATE 13, 3: PRINT "( 4 )  GENERAL (FOCAL LENGTH AND
ECCENTRICITY)"
      LOCATE 15, 3: PRINT "Enter [ _ ]"

MAINLOOP:

      LOCATE 15, 11
      RT$ = INKEY$
      IF RT$ = "0" THEN
            LOCATE 15, 11
            PRINT RT$
            CLOSE
            END
      END IF
      IF RT$ = "1" THEN
            LOCATE 15, 11
            PRINT RT$
            GOTO PARABOLA
      END IF
      IF RT$ = "2" THEN
            LOCATE 15, 11
            PRINT RT$
            GOTO ELLIPSE
      END IF
      IF RT$ = "3" THEN
            LOCATE 15, 11
```

```
                PRINT RT$
                GOTO HYPERBOLA
        END IF
        IF RT$ = "4" THEN
                LOCATE 15, 11
                PRINT RT$
                GOTO GENERAL
        END IF
        GOTO MAINLOOP

PARABOLA:
        CLS
        PRINT "WE WILL BE CALCULATING THE POINTS OF A PARABOLA"
        PRINT "Enter [0] at any prompt to terminate."
        LOCATE 5, 1
        PRINT "Enter :"
        INPUT "        REFLECTOR DIAMETER (INCHES) = ", RD
        IF RD = 0 THEN GOTO MAINMENU
        INPUT "              FOCAL LENGTH (INCHES) = ", F
        IF F = 0 THEN GOTO MAINMENU
        INPUT "                       Y INCREMENT = ", SY
        IF SY = 0 THEN GOTO MAINMENU
        LOCATE 24, 1
        PRINT "                **** GENERATING DATA FILE ****"
        MAXCOUNT = ABS(RD / 2 / SY)
        PRINT ""
        PRINT ""
        PRINT #1, "|CO-ORDINATES OF A PARABOLA WITH :"
        PRINT  , "|CO-ORDINATES OF A PARABOLA WITH :"
        PRINT #1, "|           FOCAL LENGTH  ="; F
        PRINT  , "|           FOCAL LENGTH  ="; F
        PRINT #1, "|           REFLECTOR DIA ="; RD
        PRINT  , "|           REFLECTOR DIA ="; RD

        PRINT #1, "|           F/D           ="; F / RD
        PRINT  , "|           F/D           ="; F / RD
        PRINT #1, "|    Y          X     |"
        PRINT  , "|    Y          X     |"
        PRINT #1, "|========|========|"
        PRINT  , "|========|========|"
        FOR T = 0 TO RD / 2 STEP SY
            x = T ^ 2 / (4 * F)
            PRINT #1, USING "| ###.### | ###.### |"; T; x
            PRINT  , USING "  | ###.### | ###.### |"; T; x
        NEXT T
        GOTO ANYMORE
        PRINT
        PRINT "PRESS ENTER TO RETURN TO MAIN MENU"
        INPUT RT$
        GOTO MAINMENU

ELLIPSE:
        CLS
        PRINT "WE WILL BE CALCULATING THE POINTS OF AN ELLIPSE"
        PRINT "Enter [0] at any prompt to terminate."
```

```
        LOCATE 5, 1
        PRINT "Enter :"
        INPUT "          ( A ) PARAMETER = ", A
        IF A = 0 THEN GOTO MAINMENU
        INPUT "          ( B ) PARAMETER = ", B
        IF B = 0 THEN GOTO MAINMENU
        INPUT "             Y INCREMENT = ", SY
        IF SY = 0 THEN GOTO MAINMENU
        LOCATE 24, 1
        PRINT "             **** GENERATING DATA FILE ****"
        MAXCOUNT = ABS(B / SY)

        PRINT #1, "|CO-ORDINATES OF AN ELLIPSE WITH :"
        PRINT  , " |CO-ORDINATES OF AN ELLIPSE WITH :"
        PRINT #1, "             ( A ) PARAMETER ="; A
        PRINT  , "             ( A ) PARAMETER ="; A
        PRINT #1, "             ( B ) PARAMETER ="; B
        PRINT  , " |---------------|   ( B ) PARAMETER ="; B
        PRINT #1, "|                              |"
        PRINT  , " |                              |"
        PRINT #1, "|     Y        X       X-A     |"
        PRINT  , " |     Y        X       X-A     |"
        PRINT #1, "|=========|=========|=========|"
        PRINT  , " |=========|=========|=========|"
        FOR T = 0 TO B STEP SY
            x = SQR(ABS(1 - T ^ 2 / (B ^ 2))) * A
        PRINT #1, USING "| ###.### | ###.### | ###.### |"; T; x; ABS(x - A)
        PRINT  , USING " | ###.### | ###.### | ###.### |"; T; x; ABS(x - A)

        NEXT T
        GOTO ANYMORE
        PRINT
        PRINT "PRESS ENTER TO RETURN TO MAIN MENU"
        INPUT RT$
        GOTO MAINMENU

HYPERBOLA:
        CLS
        PRINT "WE WILL BE CALCULATING THE POINTS OF A HYPERBOLA"
        PRINT "Enter [0] at any prompt to terminate."
        LOCATE 5, 1

        PRINT "Enter :"
        INPUT "                  ( A ) PARAMETER = ", A
        IF A = 0 THEN GOTO MAINMENU
        INPUT "                  ( B ) PARAMETER = ", B
        IF B = 0 THEN GOTO MAINMENU
        INPUT "        REFLECTOR DIAMETER (INCHES) = ", RD
        IF RD = 0 THEN GOTO MAINMENU
        INPUT "                     Y INCREMENT = ", SY
        IF SY = 0 THEN GOTO MAINMENU
        LOCATE 24, 1
        PRINT "             **** GENERATING DATA FILE ****"
        PRINT "                 _____"
```

```
        MAXCOUNT = RD / 2 / SY

    PRINT #1, "|CO-ORDINATES OF A HYPERBOLA WITH :"
    PRINT , "  |CO-ORDINATES OF A HYPERBOLA WITH :"
    PRINT #1, "|            REFLECTOR DIA = "; RD
    PRINT , "  |            REFLECTOR DIA = "; RD
    PRINT #1, "|         ( A ) PARAMETER ="; A; "      ( B ) PARAMETER ="; B
    PRINT , "  |         ( A ) PARAMETER ="; A; "      ( B ) PARAMETER ="; B
    PRINT #1, "|------------------|"
    PRINT , "  |------------------|"
    PRINT #1, "|    Y        X        X-A    |"
    PRINT , "  |    Y        X        X-A    |"
    PRINT #1, "|=========|=========|=========|"
    PRINT , "  |=========|=========|=========|"
    FOR T = 0 TO RD / 2 STEP SY

        x = SQR((1 + T ^ 2 / (B ^ 2))) * A
        PRINT #1, USING "| ###.### | ###.### | ###.### |"; T; x; ABS(x - A)
        PRINT , USING "  | ###.### | ###.### | ###.### |"; T; x; ABS(x - A)
    NEXT T
    GOTO ANYMORE
    PRINT
    PRINT "PRESS ENTER TO RETURN TO MAIN MENU"
    INPUT RT$
    GOTO MAINMENU

GENERAL:
    CLS
    PRINT "WE WILL BE CALCULATING THE POINTS OF A GENERAL CURVE"
    PRINT "Enter [0] at any prompt to terminate."
    LOCATE 5, 1
    PRINT "Enter :"
    INPUT "        REFLECTOR DIAMETER (INCHES) = ", RD
    IF RD = 0 THEN GOTO MAINMENU
    INPUT "              FOCAL LENGTH (INCHES) = ", F
    IF F = 0 THEN GOTO MAINMENU
GETE:
    INPUT "                      ECCENTRICITY = ", E
    IF E = 0 THEN
    PRINT " VALUE OF E = 0 IS A CIRCLE . IT IS NOT ALLOWED IN THIS PROGRAM."
    GOTO GETE
    END IF
    IF E = 1 THEN PRINT "VALUE OF 1 IS NOT ALLOWED.": GOTO GETE

    IF E > 1 THEN
        PRINT "     VALUE OF GREATER THAN 1 IS A HYPERBOLA"

    END IF
    IF E < 1 THEN
        PRINT "     VALUE OF LESS THAN 1 IS AN ELLIPSE"
```

```
        END IF

        INPUT "                          Y INCREMENT = ", SY
        IF SY = 0 THEN GOTO MAINMENU
        MAXCOUNT = ABS(RD / 2 / SY)

        A = F / (E - 1)
        C = E * A
        IF (E < 1) THEN
              B = SQR(ABS(A ^ 2 - C ^ 2))
        ELSE
              B = SQR(ABS(C ^ 2 - A ^ 2))
        END IF
        PRINT , "    **** GENERATING DATA FILES ****"
        PRINT , ""

        PRINT #1, "|CO-ORDINATES OF A CURVE WITH :"
        PRINT  , "  |CO-ORDINATES OF A CURVE WITH :"

        PRINT #1, "|              REFLECTOR DIA ="; RD
        PRINT  , "  |              REFLECTOR DIA ="; RD
        PRINT #1, "|    (A)="; A; "    (B)="; B; "    (C)="; C; "    (E)="; E; "   (F)="; F
        PRINT  , "  |    (A)="; A; "    (B)="; B; "    (C)="; C; "    (E)="; E; "   (F)="; F

        PRINT #1, "|—————————————————|"
        PRINT  , "  |—————————————————|"
        PRINT #1, "|     Y          X          X-A    |"
        PRINT  , "  |     Y          X          X-A    |"
        PRINT #1, "|==========|==========|==========|"
        PRINT  , "  |==========|==========|==========|"

        FOR T = 0 TO RD / 2 STEP SY
              IF (E < 1) THEN
                    x = SQR(ABS(1 - T ^ 2 / (B ^ 2))) * A
              ELSE
                    x = SQR(ABS(1 + T ^ 2 / (B ^ 2))) * A
              END IF
              PRINT #1, USING "| ####.### | ####.### | ####.### |"; T; x; ABS(x - A)
              PRINT  , USING "  | ####.### | ####.### | ####.### |"; T; x; ABS(x - A)
        NEXT T
        GOTO ANYMORE
        PRINT
        PRINT "PRESS ENTER TO RETURN TO MAIN MENU"
        INPUT RT$
        GOTO MAINMENU

ANYMORE:
        CLOSE #1
        PRINT
        PRINT "DATA FILE HAS BEEN GENERATED"
```

```
            PRINT
            PRINT "RE-FORMATTING ("; MAXCOUNT; ") ENTRIES"
            REDIM A$(4 * MAXCOUNT)
            DIM HEADER$(6)
            OPEN "I", #1, "DATA1.$$$"
            FOR I = 1 TO 6
                    INPUT #1, HEADER$(I)
            NEXT I
            N = 1
REFORMAT:
            IF EOF(1) THEN GOTO READDONE
            INPUT #1, A$(N)
            N = N + 1
            GOTO REFORMAT
READDONE:
            OPEN "O", #2, "DATA2.$$$"
            GOSUB WHEADER
            PRINT #2, HEADER$(5), HEADER$(5)
            PRINT #2, HEADER$(6), HEADER$(6)
            OS = N \ 2
            FOR W = 1 TO OS
                    PRINT #2, A$(W), A$(W + OS)
            NEXT W
            CLOSE #2'

SAVEIT:
            PRINT " WOULD YOU LIKE TO SAVE THIS DATA FILE (Y/N) "
            INPUT "", YN$
            IF YN$ = "Y" OR YN$ = "y" THEN
                    INPUT "ENTER FILENAME : ", FILENAME$
                    SHELL "COPY DATA2.$$$ " + FILENAME$
            END IF
            PRINT "WOULD YOU LIKE TO PRINT-OUT THIS DATA FILE (Y/N) ";
            INPUT "", YN$
            IF YN$ = "Y" OR YN$ = "y" THEN
                    SHELL "COPY DATA2.$$$ LPT1:"
            END IF

            PRINT "WOULD YOU LIKE TO RUN ANOTHER DATA FILE (Y/N) ";
            INPUT "", YN$
            IF YN$ = "Y" OR YN$ = "y" THEN GOTO MAINMENU
            END

WHEADER:
            FOR I = 1 TO 4
                PRINT #2, HEADER$(I)
            NEXT I
            RETURN
```